c|net Do-It-Yourself Camera & Music Phone PROJECTS

24 cool things you didn't know you could do!

Ari Hakkarainen

New York Chicago San Francisco
Lisbon London Madrid Mexico City
Milan New Delhi San Juan
Seoul Singapore Sidney Toronto

The McGraw·Hill Companies

McGraw-Hill books are available at special quantity discounts to use as premiums and sales promotions, or for use in corporate training programs. For more information, please write to the Director of Special Sales, Professional Publishing, McGraw-Hill, Two Penn Plaza, New York, NY 10121-2298. Or contact your local bookstore.

CNET Do-It-Yourself Camera and Music Phone Projects:
24 Cool Things You Didn't Know You Could Do!

1234567890 QPD QPD 019876

ISBN-13: 978-0-07-148564-7
ISBN-10: 0-07-148564-3

Sponsoring Editor
Judy Bass

Editorial Supervisor
Jody McKenzie

Project Editor/Manager
Vasundhara Sawhney

Editorial Assistant
Laura Hahn

Copy Editor
Lucy Mullins

Proofreader
Raina Trivedi

Indexer
Robert Swanson

Production Supervisor
Jim Kussow

Composition
International Typesetting and Composition

Illustration
International Typesetting and Composition

Cover/Series Design
Jeff Weeks

Cover Illustration
Sarah Howell

Contents

Part II Challenging

Foreword

It wasn't long ago that they laughed. "A camera in a cell phone? That's crazy."

Nobody laughs anymore. The cell phone seems to gobble up other technologies like Pac Man, integrating them into its svelte, indispensable little frame. Today it's hard to find cell phones that *only* make calls. From taking pictures to reading e-mail, these features are there because we love them, and because the best portable technology is the one you always have with you.

The latest cell phone revolution is putting music, video, and TV broadcasts in our phones, along with Web and e-mail access. It's no longer absurd to ask if the phone will one day replace our PC, digital camera, and PDA.

Still, it always surprises me how many people are carrying around a phone that includes a digital camera, a Web browser, an e-mail terminal, and a digital organizer, and yet all they do is make calls. They know their phone does more, but nobody has shown them how to take advantage of these abilities. In this book we address that with fun, useful projects selected to answer the most common "how do I..." questions we get here at CNET. Dig in and get to know the rest of your amazing cell phone.

Brian Cooley
CNET Editor-at-Large

Acknowledgments

Too often, I take the services, information and applications that are available on the Internet as natural as the salt in the sea. Probably I'm not the only one who tends to forget how dramatically access to information and communication services has improved in the last 10 years because of the Internet and mobile networks. I have used applications and services in this book that didn't cost me a cent and won't cost anything for you either. I'd like to thank the individuals behind those services and applications, but since I don't know their names, I simply thank the creators of Anvil Studio, Doppler, IrfanView, MidiPiano, Mobilcast, Open Media Network, Opera, Orb, Podcasting, S60 Internet Radio, ShoZu, XnView, and the community that has built the Typo3 system.

I have made it my job to share information. It's not only about delivering advice to the readers, but also listening to them. I'd like to thank all the readers of Avec Mobile web site. They have kept me focused on what is important and what is hollow hype in modern technology. New and old readers, please share your thoughts, join the discussion and discover extra tips related to this book at www.avecmobile.com/book. You may also contact me directly at: book@avecmobile.com.

Thanks to the professional people at McGraw Hill who trusted me with these 24 projects, you are now reading one of the first books that explains how you can use your modern phone for digital photography, video, and music.

It's about five years since I and other members of the family completely switched to mobile phones for telephone service. Liisa and Leo may not always be as enthusiastic about new technology as I am, but I always appreciate a phone call, a message, or a photo from them. Always.

About the Author

Ari Hakkarainen of Helsinki, Finland, has more than 20 years' experience in the high tech industry. He is the creator and operator of avecmobile.com, an online publication that provides practical advice to consumers on a wide range of portable technology devices, such as camera phones, smartphones, digital cameras, MP3 players, GPS navigators and related software.

Introduction

Why on Earth would anyone want to listen to music on a phone? Who would like to watch TV on the phone, or take photographs and send them to the Internet for the whole world to see? Aren't phones meant for talking with other people?

Cell phones, mobile phones, or hand phones (I simply call them phones from now on) are used for many other things besides talking with other people. The world has already entered a new era when lots of useful and fun tasks, such as taking pictures, watching movies, listening to music, posting new entries on a blog, or uploading photos from the phone to a photo album are available for anyone with a modern phone. But before all that, you need to learn a few things about these new devices. Let me tell you how I learned my lesson.

A few years ago, when I had my first camera phone tucked in the pocket for the second day in a row, I headed to the mall for weekend grocery shopping. I walked along a long shelf of cheese products when my eye caught a familiar face behind the sales desk. A celebrity, winner of several beauty contests, was standing behind the desk, introducing new cheese flavors to shoppers. Undisturbed, I continued to the cashier and took my place in the queue. Only then, I realized that I had a camera with me. I left my place in the queue, walked back to her, whipped out my camera phone, asked her permission to photograph and snapped a picture of her.

When I got out of the mall and sat down on the front seat of my car, I realized another important thing. Not only I had a camera with me, but I could send the captured photo to my friends right away. So, in my car at the parking lot, I composed a message, attached the photo to it and sent it to my colleagues (I knew they had camera phones already then). I expected furious applause and envious remarks about the subject of the picture, but all I got was total silence. On Monday, in the office, I asked my colleagues if they had received the photo message. Yes, they had received it, but they couldn't figure out what the photo was all about.

I learned my lesson. I realized I had to apply my knowledge about mobile technology and photography to this new product. I also thought that many people would avoid the same mistakes if I would tell them how (if you want to know, I was too far away from the subject and the lighting was poor—the image was unbelievably fuzzy). This experience was one of the sparks that led to the creation of the Avec Mobile web site.

If you already own a camera phone or a smartphone, or if you are considering buying one, you can learn a lot from this book. It doesn't matter if your phone happens to be a BenQ, HTC, LG, Motorola, Nokia, Samsung, Sanyo, Siemens, Sony Ericsson, or another popular brand. It doesn't matter who your service provider is. Independent of the brand, modern phones come with many common features, such as memory card,

Bluetooth, MP3 ringtones, or possibility to download music and upload videos. You can learn to take advantage of those features. That's why this book is in your hands.

You'll find projects in this book that show you how to correctly expose photos, copy information between a PC and a phone, and download songs and ringtones to the phone, to name a few projects. In addition, many projects include detailed instructions for phones powered by the Windows Mobile and the Symbian OS / S60 software.

A friend of mine lives about a 90 minute-drive away from my house. After she had driven the distance for the first time, she told me that the radio and the CD player in her car were not working. "How can you drive for 90 minutes and only listen to the engine and faster cars whiz by?" I asked. "I listened to the radio that I have on my phone", she said. "I've been using it for quite awhile and you know what? If someone calls me while I'm driving, I already have my headset connected and the music automatically mutes when I answer the phone."

Let new mobile devices make your life a little bit easier and more fun. Here's how.

Modern Phone Is a Powerful Multimedia Device

Your new phone can be a digital camera, portable audio player, email reader, camcorder, GPS navigator, Internet access device, and a couple of other useful things. It's quite a bundle in a small handheld device. Perhaps that's the reason why we have so many names for them, such as camera phone, camcorder phone, music phone, MP3 phone, or smartphone. Let's find out what exactly are they.

Camera Phone

Camera phone is a phone with a built-in digital camera (Figure 1). A small lens at the back panel of a phone, or at the cover of a flip phone is often the only visible element indicating the phone is also a digital camera. Practically every camera phone can record video clips as well. Products that can capture relatively high quality video are often called camcorder phones.

Figure 1

A camera phone (Motorola MPx220) with a dedicated camera button and a memory card slot (protected by a rubber cover) on the side.

Taking a photograph on a camera phone is simple. Activate the camera from the phone's menu system, push a dedicated camera key, or remove the lens cover, and you are ready to shoot. When you take a photo on a camera phone, it is saved in the phone's internal memory, or on a removable memory card. Also digital cameras use memory cards as their storage media.

You can copy the saved pictures and videos from the phone to a computer. Camera phones that come with Bluetooth or infrared connectivity let you transmit information to computers and other phones wirelessly. A memory card or a data cable is often used for transferring large amounts of information between a computer and a phone.

One of the best things with camera phones is that you can instantly share your master shots. You may send the images to other phones, to email addresses, to a shared photo album, or video sharing sites on the Internet. You need a data communication plan for your phone from your service provider in order to access email or the Internet.

Music Phone

Monotonously beeping ringtones were the humble beginnings for music phones. Next, phones were able to play polyphonic MIDI tunes and finally, real music (Figure 2). Now, music phones can play the same MP3 tracks that your computer or iPod can play. Podcasts (downloadable talk shows) can also be played on a music phone. One of the best things with music phones is that you don't have to worry about missing a phone call when listening to music, because the music will automatically pause when the phone rings.

Figure 2

A music phone (Nokia N91) with large memory capacity for storing music, audio books and podcasts. Dedicated music keys, volume control keys and Hi-Fi headphone interface allow for easy access to the music.

When you want to listen to music on your phone, you launch the music player application, select tracks and push Play. Simple as that. Before that, however, you must have copied the tracks to the device. A good place to start looking for music to download is your own CD library. You can copy your CDs to the computer and then transfer the songs to the phone. It is also possible to download MP3 files from the Internet directly to the phone.

Some music phones come with an FM radio. The phone headset doubles as a radio antenna and you tune into radio stations by accessing a radio tuner application on the phone. Even if there's no FM radio on the device, you maybe able to download an application for tuning into Internet radio stations.

If it is possible to store a few CDs worth of MP3 music on a phone and listen to the tunes on high-quality headphones, why would anyone need a dedicated MP3 player or iPod anymore? A dedicated MP3 player can still produce better audio quality for hi-fi minded music lovers, and they typically come with larger storage capacity than music phones.

Smartphone

When the features of a mobile phone and an PDA (personal digital assistant) are integrated into one product, the result is a smartphone (Figure 3). The extensive set of features in smartphones includes Internet access, email, calendar, data synchronization, document viewing, music, photos, video, GPS navigation and a few more things. Smartphones often come with both camera phone and music phone features.

Figure 3

A smartphone (Qtek/HTC 8300) may look like another phone, but there is power inside the device. The operating system software makes the device smart—in this case, it's Windows Mobile.

What really makes smartphone smart is the software inside the device. Windows Mobile, Symbian OS, Palm OS, and Linux are examples of operating systems that are powering smartphones. These operating systems come with documentation and tools that makes it possible for anyone who can write software to create applications for these devices. Then we - smartphone owners - can download these applications and enhance the functionality of our mobile devices.

One of the best things with smartphones is that when you learn to use one, you can very easily use any smartphone that runs on the same operating system software regardless of its hardware brand.

Identify the Software That Is Powering Your Phone

A rapidly increasing number of phones and other mobile devices are powered by an advanced operating system software, such as the Symbian OS, Windows Mobile, Palm OS, or Linux. Although some of the product names are familiar for computer users, these software products have specifically been designed to run on small battery-powered devices. Phones built on the advanced operating systems tend to be at mid-range or high-end price categories.

On the other hand, many mid-range and low-end phone models are powered by manufacturers' own operating system software. Each one of these products look different, have different features and behave differently.

Identify the Operating System on Your Phone

note *An operating system is a fairly large and complex piece of software that makes your phone tick. Among other things, it displays the menus for you, plays the ringtones, encodes and decodes your phone calls, and sends and receives the bits that compose your messages. Regardless of the type and brand of your phone, there is an operating system inside the device. It is the same thing as the Windows XP on your PC or the OS X on your Mac, although you usually can't upgrade or change the operating system on your phone.*

It may not be as self-evident as on a PC which operating system a phone is running. Take a close look at the unit for signs of branding. For example, Windows devices come with a small 'Designed for Windows Mobile' logo at the front or at the back of the unit.

tip *You can identify your phone model even if you have lost your product box and user guide. Open the unit's battery cover and remove the battery. The model number is printed on the product identification label (Figure 4).*

Figure 4

Discover the phone model information under the battery. In this case, the model is MPx220.

The product box, user guide, or manufacturer's web pages often specify the name and version of the operating system software. If you can't find the information, browse the following web pages until you discover your phone model.

Windows Mobile View the following web page and see if you can spot your device: www.microsoft.com/windowsmobile/smartphone.

If you identified your phone as a Windows Mobile device, you'll be frequently using the Home key and Home screen. The Home key is a dedicated button on the keypad with a picture of a small house on it (Figure 5). The size and the placement of the key may vary, but you'll always find it on the keypad. You can push the key whenever you want to. Even when you are taking pictures, or listening to music, you may hit the Home key. The application that you were using continues to run in the background.

Figure 5

The key with the picture of a house on it represents the Home key on a Windows Mobile phone.

When you push the Home key, you'll see the Home screen (Figure 6). At the top of the screen, you can see icons for applications that are already running. You can move up and down, left and right with the arrow keys or with the navigation key. At the bottom left corner of the screen, you can see the Start key. Hit the Start key when you want to go to the menu system for viewing all the applications, photos, music, tools and settings of the phone.

Figure 6

The Home screen on a Windows Mobile phone.

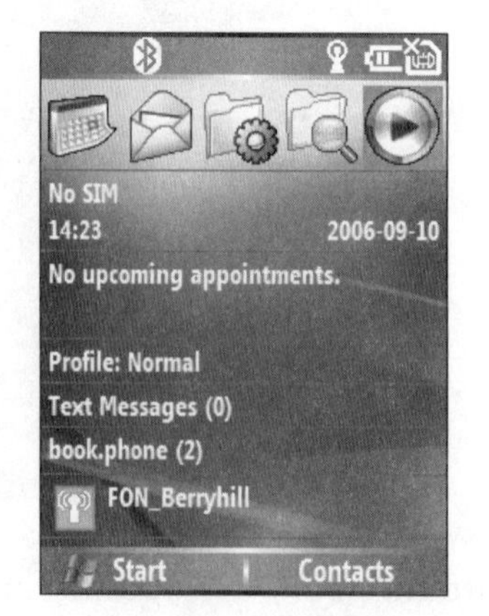

Symbian OS To find out if your phone is powered by the Symbian OS, go to the following web page: www.symbian.com/phones.

Pay attention to the identifier S60, UIQ, or MOAP that is associated with the phone model. These abbreviations represent different menu systems (user interfaces) that run on the Symbian OS software. The user interfaces look and behave slightly differently. More details about S60 (also known as the Series 60 Platform) devices can be found on the www.s60.com/devices web page. More information about UIQ devices is available at www.uiq.com/uiqphones. Phones with MOAP user interface are available in the NTT Docomo network in Japan.

If you identified your phone as a Symbian OS / S60 device, you will often need the Menu key (Figure 7) and the main Menu in the projects discussed in this book.

Figure 7

In this phone, the Menu key (two lines with circles at the opposite ends) is located at the center of the keypad, just below the large selection/navigation key.

The Menu key has two functions. You can toggle between the main Menu and the Standby screen, and you can access the applications that are running in the background. The main Menu (Figure 8) is your gateway to all the features of the phone. You can find all applications, settings, photos, music and videos by navigating the folders in the main Menu.

Figure 8

The main Menu on a Symbian OS / S60 phone.

The Standby screen is sometimes called the idle screen or the home screen (Figure 9). In any case, it is the screen where you can see items like the date and the time, number of new messages, and missed phone calls.

Push the Menu key and hold it down. A small window will pop up displaying the applications running in the background (Figure 10). You can switch to any application by selecting it from the list.

Figure 9

The standby screen on a Symbian OS / S60 device.

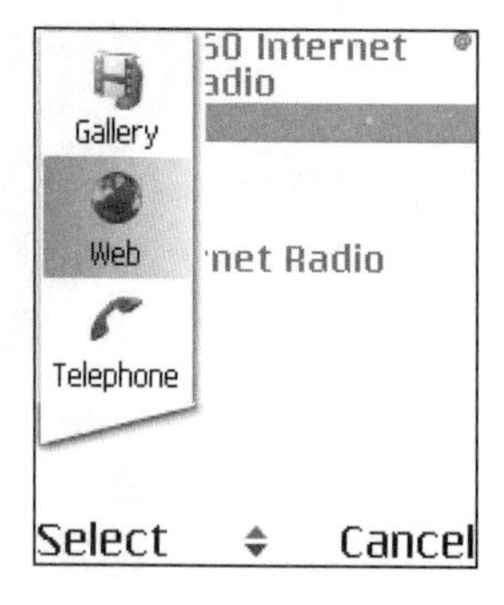

Figure 10

Switch to an application running in the background.

Palm OS You can find information about Palm OS mobile devices here: www.palmsource.com/products.

Linux Information about devices that are running on Linux is available here: www.linuxdevices.com.

Other phones Other phones than the ones powered by the Windows Mobile, Symbian OS, Palm OS, or Linux are likely to run an operating system developed by the device manufacturer.

You can easily follow the projects in this book even if your phone isn't powered by the Windows Mobile or Symbian OS. Just make yourself familiar with your phone's features. For instance, examine the menus associated with the digital camera, or find out where the email settings are located. Then, you can learn, for example, how to copy photos from your phone to a PC, create a new ringtone on the PC download it to your phone, or upload a video to a video sharing site regardless of the type of phone you have.

Identify the Essential Features on Your Phone

New camera phones and music phones may have hundreds of features. Some of them are vital for digital photography and some are necessary for playing music. Use the following table as a reference for features that are required for successful photography and for features required for enjoying music on the phone.

The abbreviations Y, N and U mean the following:
Y = Yes, you need this feature for photography or music.
N = No, you don't need this feature for photography or music.
U = Useful but not absolutely required feature.

Feature	Required for digital photography	Required for playing music
Color screen	Y	U
Digital camera	Y	N
Flash	U	N
Video recording	Y	N
Printing directly from the phone	U	N
Removable memory card	Y (1)	Y (1)
Data cable	U	U
Bluetooth	U	U
Infrared	N	N
MP3 music player software	N	Y
MP3 songs as ringtones	N	Y
FM radio	N	U
Stereo headphones	N	Y
3G (EDGE, EV-DO, HSDPA, or UMTS/WCDMA) network connectivity	U (2)	U (2)
Wi-Fi connectivity	U	U
E-mail software	Y	U
MMS (Multimedia messaging)	U	N
Web browser software (HTML or XHTML)	Y	U

1. A removable memory card is required for storing photos and music, unless there is another storage media in the device, such as a hard disk.
2. Either 3G or Wi-Fi connectivity is required for downloading and uploading information and accessing the Internet.

Digital Photography and Music Features Exposed

Color screen Color screen (instead of a black-and-white screen) is essential on a device that's used for digital photography. A high-resolution screen lets you view sharp images of photos you have captured. A bright screen is useful when taking photos in the daylight, because the phone screen doubles as a viewfinder for the camera.

Screen resolutions from 176x208 pixels up to 640x480 pixels are all suitable for photography. A camera phone screen should be able to display at least 65000 colors. Flip phones often come with two screens: a small screen on the cover and a larger screen inside the unit. The small screen is primarily intended for viewing the phone number or the name of the caller without having to open the phone.

Digital camera The size of a digital image is measured in pixels (picture elements). Image resolutions in camera phones start from 0.3 megapixels (640x480 pixels, or VGA). Many products can capture 2 megapixel (1600x1200) or higher resolution images. Megapixels, however, measure the image size, not the quality. The difference in image quality between the best and worst camera phones can be dramatic. The best way to find out about image quality is to compare photos taken on different camera phones.

0.3 megapixel camera resolution is sufficient if you only intend to send photos to other small-screen devices. If you want to print your pictures, or view them on a computer monitor or on a TV, you should use one megapixel (about 1280x960 pixels), or higher resolution camera.

Some camera phones come with two lenses: one at the back (Figure 11) and another at the front. The lens at the back is for photography. The lens at the front panel is intended for video calls. It streams live images of the person holding the phone to the other end of the line. Although it is possible to capture still images using the front lens, the quality tends to be poor.

Figure 11

Digital camera lens and flash located at the top of the cover.

Flash Flash in camera phones is often placed right beside the lens (Figure 11). A typical flash sheds light at an arms length from the photographer, which is usually not enough for capturing decent indoors photos. Check the phone's camera menus for flash controls that allow you to switch it on and off.

Video recording Most camera phones can capture video clips. Typical resolutions for camera phone videos (even on a megapixel model) are 176x144 (QCIF) and 320x240 (QVGA) pixels. The 176x144 resolution is fine for watching a movie on the phone. If you intend to watch your camera phone videos on a computer or on a TV, 320x240 is the minimum resolution. You can find the video recorder from the phone when you activate the camera and browse the menus for video mode.

Memory card Removable memory card (Figure 12) is the most versatile media for storing photos, videos and music on a portable battery-powered device. The memory card makes another important thing possible: transfer of photos, music and other

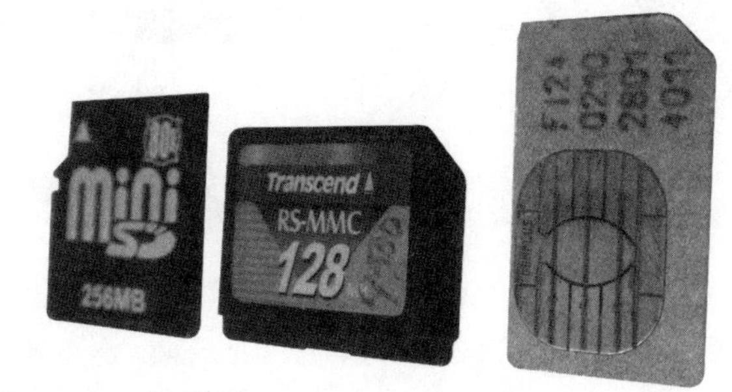

Figure 12
Mini SD and RS-MMC are widely used memory cards on mobile devices. The phone service subscriber identity module, SIM card, on the right is for size comparison only.

items between your phone and computer. For example, you can print your photos by removing the memory card from the phone and inserting it into a printer. Also, you can copy an unlimited number of ringtones from your computer to the memory card and use them on your phone whenever you want.

Although many types of memory cards are available, it doesn't really matter which type of card your phone happens to accept—MMC, SD, Memorystick, or something else. Get a low-cost memory card reader if your computer doesn't have a built-in memory card reader. Connect the external card reader to your computer's USB port and you are ready to exchange information.

The memory card slot is usually located on a side, or at the back of the unit. In some phone models, memory cards have even been hiding under the battery. If your phone didn't come with a memory card slot, use Bluetooth, infrared or a data cable for copying information between your phone and computer.

Printing If you want to print photos directly from the phone without copying them to a computer, you must have a printer that can communicate with the phone. PictBridge, a communication technology developed by camera and printer manufacturers, provides a standard communication method for printing. You need wireless Bluetooth connection or a USB cable for sending the print job from the phone to a printer. Search for Printer or PictBridge entries in the phone menus. It is also possible to download printing applications for some smartphones.

Bluetooth and infrared Bluetooth and infrared wireless communication technologies have been designed to transmit information between phones, between a phone and a computer, or between a phone and other devices. You can copy photos, music and other information via Bluetooth (the signal travels over radio waves) or via infrared (over light waves). Infrared is increasingly being replaced by Bluetooth, because it doesn't require line of sight during the transmission.

If you can choose between infrared and Bluetooth, take Bluetooth (Figure 13). It is more reliable communication technology and it can be used with wireless headsets as well. Check the connectivity settings on your phone for activating Bluetooth.

Data cable Moving large amounts of information via data cable is faster than via Bluetooth or infrared connection. If you can choose, USB cable is the best choice. Often, the data cable is included in the product package or you can purchase the data cable as an accessory.

Figure 13

Scanning for Bluetooth devices. Bluetooth and infrared are managed in the same way on this phone.

ss000.bmp
Select 'Beam' to send.
IR
ROUBAIX
Linramin
Zokvio
Searching
Beam
Cancel

MP3 player software MP3 player software, built-in loudspeakers and stereo headphones are the elements that make phone a portable music device. Typically, the MP3 applications can play other types of digital music, such as AAC, or WMA as well. Keep in mind that songs purchased from online shops, such as the iTunes Store or Napster, are compatible with a few select mobile devices only.

A good music phone can play songs while you do something else, for instance, view photos. Playing music requires a fair amount of power. Long periods of music listening dries up the battery faster than normal use. You can find the music player from the phone menus, or you maybe able to manage the music from the dedicated keypad (Figure 14).

Figure 14

Dedicated music control keys on the keypad.

FM radio An FM radio on the phone consists of an antenna embedded in the headset cable and of a tuner software. The tuner software is an application that you can find from the phone menus. If you find it, hook up the headset—it must be connected for listening to the radio.

Stereo headphones Phones designed for music come with stereo sound and stereo headphones (Figure 15). You may occasionally want to plug in ordinary headphones for better sound. If the jack on the phone is not a standard headphone connector, there should be an adapter included with the phone. In daily use, the included headphones are practical because of the remote control that lets you adjust the volume and answer phone calls without having to pick up the phone from the pocket or bag.

3G connectivity 3G (Third Generation mobile) networks let you access the Internet and email from your phone at broadband speed. Common 3G network technologies

Figure 15

A stereo headset with remote control.

are EV-DO, HSDPA, UMTS/WCDMA, and broadly speaking, EDGE as well. EDGE is somewhat slower than the 3G network technologies whose theoretical speeds are measured in megabits per second. In practice, information is transmitted at about 256-512 kbps (kilobits per second). You need 3G (or EDGE) connectivity when you post pictures on photo and video sharing sites, access email, download music, or watch your home TV on the phone.

In most countries of the world, you don't have to choose between 3G technologies because all service providers use the same technology. Usually, they offer different services at different speeds and price. In countries where you can choose between 3G technologies, choose the one that works in the area where you live and travel within your country and overseas. If you are shopping for a new phone, make sure it comes with 3G or Wi-Fi connectivity.

Wi-Fi connectivity Wi-Fi wireless technology is a popular method for connecting a notebook computer to a local area network and to the Internet. Exactly the same technology is available for many new phones. In theory, Wi-Fi is extremely fast (11–54 Mbps), but it is up to your phone how fast it can really push bits through. In practice, today's phones transmit data over Wi-Fi at speeds from 512Kbps to 1Mbps.

Wi-Fi is a convenient connectivity option for phone users because you can connect to an existing Wi-Fi access point at home or in the office, without having to pay extra for the access. Wi-Fi requires more power than other communication technologies so you have to monitor the battery closely when you are using it (Figure 16).

Figure 16

Save battery power by switching Wi-Fi on only when you need it.

Email Despite the relatively small screen and small keypad, email is one of the most useful features on a phone (Figure 17). The possibility to send recently snapped photos to a friend, post a blog entry, or upload a home movie on a video sharing site from the phone simply makes life more fun. Search for email in the messaging or communication menus in your phone.

Figure 17
Access the same email account on your phone as you do on your computer.

MMS (Multimedia Messaging Service) MMS makes it possible to send photo messages from phone to phone. It is possible to include audio and video in MMS messages as well. In addition to sending MMS messages to phone numbers, you can transmit MMS to email addresses. However, you can only receive MMS that has been sent as an MMS message, not as an email. If you have a camera phone, the chances are that you have MMS capability on your phone.

Web browser Internet Explorer, Firefox, Safari, Opera and Netscape are familiar applications for computer users, because they bring web pages from the Internet to our computer monitors. Despite the small screens, many phones can display exactly the same information as computers, only the layout is different (Figure 18). Phones that can do this have HTML/XHTML browsers. This capability is valuable because then you can access, for example, music, photo sharing sites and blogs on the Internet from your phone. Avoid phones that rely on the limited version of HTML, WAP.

Figure 18
Internet Explorer on a phone.

Part I

Easy

Project 1

Take Better Pictures with Your Camera Phone

What You'll Need:

- **A memory card for storing pictures**
- **Cost: $10 or more, depending on the card's storage capacity**

When you push the shutter button on your camera phone, you hear a click and the device captures a digital image. Simple as that, right? Yes, but snapping sharp and well-exposed pictures requires some practice. In fact, there are many little things that are easy to learn and will improve your chances of getting more memorable photos with your camera phone.

Step 1: Keep the Phone Steady

One of the most common reasons for blurry photos is a slight movement of the lens when the image is being captured. Often, the phone moves when you push the shutter button, or you believe that the picture already has been saved and you stash the phone away, when in fact, the camera lens is still letting light through to the image sensor. The less light there is, the longer the lens must let light through to the image sensor and the longer you have to keep the phone steady.

For steadying the phone, support it with both hands or hold it against a tree, traffic sign or any solid object. If you can't find anything to support your hands, hold your elbows against your body. It will stabilize your hands.

If you are still shopping for a camera phone, consider a model that has a dedicated camera shutter button (see Figure 1-1). The dedicated camera key is usually placed in a position that gives you a good and comfortable grip when you hold the unit in your hand.

Figure 1-1

A dedicated shutter button is better for photography than a multifunction key.

Step 2: Use Digital Zoom with Caution

Digital zoom and optical zoom are two totally different things. Optical zoom can be a big help when framing photos, because the camera optics brings objects closer by magnifying them. The use of optical zoom doesn't directly affect the image quality or sharpness.

Only a few camera phones come with an optical zoom, but many are equipped with digital zoom. When you use digital zoom, objects appear to get closer as if you were using optical zoom, but that's not the case. The digital zoom only magnifies the image in the camera software, degrading image quality in the process (see Figures 1-2 and 1-3). Avoid using digital zoom altogether and use your feet to get closer to the object.

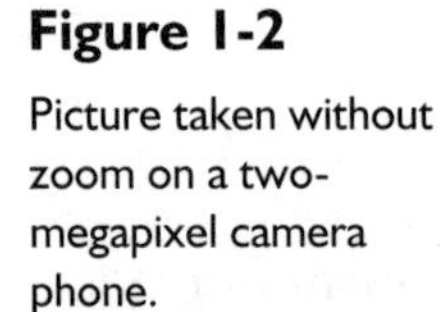

Figure 1-2

Picture taken without zoom on a two-megapixel camera phone.

Figure 1-3
The same picture as in Figure 1-2 but taken with 10× digital zoom applied.

Step 3: Use Flash for Indoor Photos

When you take photos indoors, use the flash or find a location with plenty of light. Nothing will save you from murky, unrecognizable faces if there isn't enough light for the lens. If you have a new phone model that came with a built-in flash, use it (see Figure 1-4). The average camera phone flash won't do miracles in poor light, so you should take a test picture to verify the image quality. Frequent use of the flash will dry your battery, but photos are worth it—they last longer than the camera or the battery.

Figure 1-4
Picture taken with a flash on a two-megapixel camera phone three feet (one meter) away from the subject.

tip

If there's no flash in the camera and your test picture is too dark, switch to Indoors or Night Scene mode. You should find it when you activate the camera and open the camera menu. If the Night Scene mode doesn't help, or if there's no such function on your camera, you have to be creative. Look for extra light. For instance, find lamps that could shed more light on the subject, find alternative locations where you could move the subject, or try to shoot from another angle.

tip

Adjusting the color balance, or white balance, is another technique for improving image quality for photos taken indoors. You may have noticed that sometimes photos taken in lamplight don't seem to have natural colors. There's nothing wrong with your camera phone. The effect is created because different light sources, for example, the sun or fluorescent lights, produce different wavelengths. Most camera phones automatically try to adjust the color balance for the light, but sometimes manual setting is required. You should find the White Balance menu selection when you activate the camera and open the menu. Look for Color Balance or White Balance. Then choose an entry for the light source that illuminates the room.

Step 4: Side Light Gives Good Contrast for the Image

When you are photographing outdoors, and you can take advantage of the daylight, it is possible to take excellent photos almost anywhere. Still, avoid situations where the sun is shining straight into the lens, because the image sensor can be exposed to excessive light and the whole frame may be washed out. If the light is reflected from behind the camera and straight into the object, the object may lack contrast.

The safest technique is to position yourself so that the object is lit from a side. This way, you get enough light and nice contrast for the image.

Step 5: Focus on One Object

Photos rarely do justice to wide-open scenery, no matter how beautiful the landscape happens to be. Even with a high-end digital camera, it is difficult to capture the depth and the space of a landscape. Especially with a camera phone, it is better to look around and try to find an interesting object that could be portrayed in the foreground. Then, try to find a camera angle so that the beautiful scenery is in the background.

Step 6: Pay Attention to Framing Your Shots

If you give your camera to someone who isn't into photography and ask the person to take your picture, it is very likely that the person will position you in the center of the frame. This usually results in a somewhat dull photo (unless you are making funny faces).

A simple rule for framing a picture is to divide the image area into three horizontal and three vertical sections. This gives you a grid of nine boxes inside the frame (see Figure 1-5).

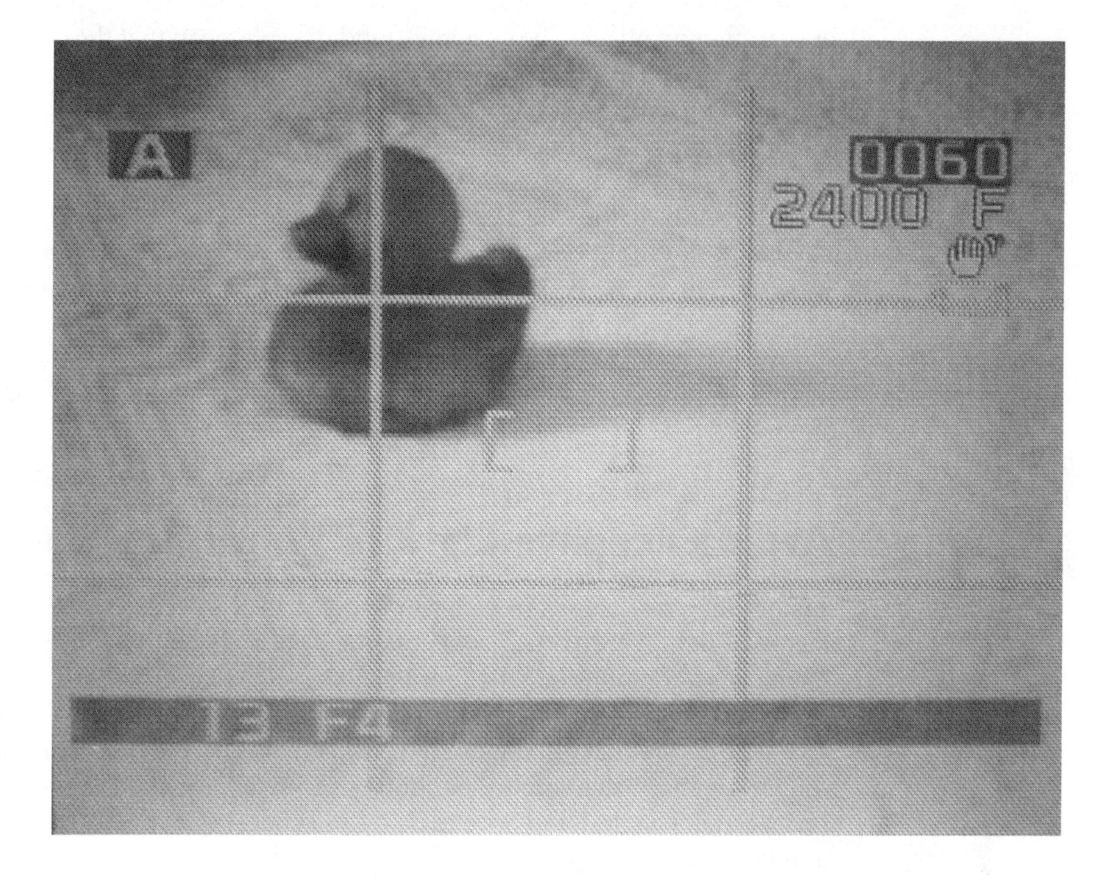

Figure 1-5
Framing a shot with the grid.

The secret for balanced photos is to position the subject into one of the intersections on the grid. These intersections, the corners of the center box in the frame, are the anchor points for the main topic of your image.

Step 7: Try Different Camera Angles for More Dynamic Images

You can make your photos more dynamic by experimenting with different camera angles. For example, point the lens to the subject from a lower or higher angle than you would normally do. Hold the camera at some other level than your eye level. If there are people blocking your view, raise the camera above your head and point it to the subject. You'll need to take several photos to get one that's worth saving, but it may be your master shot.

Step 8: Timing Is Everything When Shooting Moving Objects

There is a delay between the moment you press the shutter button and when the camera captures an image. The clicking sound doesn't mean that the camera has captured the photo; it just indicates that you pressed the shutter button.

If you push the shutter button when a moving object already has reached the center of the frame, it is quite likely that the object will have disappeared from the frame before the camera captures the image.

It depends on the direction and the speed of the object how early in advance you have to press the shutter. If you are close to the object or if it is moving fast and sideways in front of you, you have to be well prepared and push the shutter early enough, sometimes even before the object enters the frame.

Another technique for framing a moving object is to smoothly follow the movement with the lens. It is important to follow the object before and after you press the shutter.

Step 9: Avoid Excessive Light in Water and Snow

Taking photos of objects in water or snow is tricky, because water and snow reflect plenty of light. When an object is photographed against water or snow, it easily remains as a dark, unrecognizable entity.

Try to point the lens on different spots on and off the object in order to find an area that doesn't have such a high contrast between the dark and light areas in the frame. Another technique for avoiding excessive light from the water or snow is to find a camera angle—you have to be creative here—that doesn't reflect so much light into the lens.

Step 10: Take Extra Photos

Whenever you are taking pictures and the situation allows it, take a couple of extra photos. A camera phone is an excellent device for capturing all kinds of photos, but in the heat of the moment, it may be impossible to tell if the image you shot was a success or not.

Some people claim that the first photo on any subject is always their best, and they rarely take any backup shots. Still, take the extra photos; you can easily remove the failed pictures later.

Project 2

Copy Photos and Other Information from Your Phone to a PC Using a Memory Card

What You'll Need:

- **A memory card. It doesn't matter which type the card is, as long as it is compatible with your phone**
- **A memory card reader. If you already have a memory card reader for your digital camera, check if it can read the memory card used in your phone. Also, many new laptop computers come with built-in memory card readers**
- **A USB port in your computer for the memory card reader. Practically all computers manufactured in this century have at least one USB port. If there's no USB port on your computer, you can install a USB interface card inside your computer**
- **Cost: $10–$50 U.S. for a memory card, dependng on its storage capacity**

  $15–$30 U.S. for a memory card reader

  $40–$60 U.S. for a USB interface card

Inevitably, the moment will arrive when the storage space on your phone becomes full. Don't simply remove your valuable memories to make space for new photos! Instead, move the photos from the phone to your computer where you can organize them into albums, share them, print them out, and back them up on CDs.

An easy and fast way to copy all photos (along with other information) to a computer is to use a memory card as the transfer medium. A memory card allows you to transfer any information: photos, video clips, MP3 music, software applications, and messages. The best thing about the memory card is that it works both ways: you can also copy MP3 music, wallpapers, ringtones, and software *from* your computer *to* your phone.

note *There are other techniques for copying information from your phone to a safe storage; you can off-load your photos by sending them to an online photo album or beaming them over Bluetooth or infrared to another device or computer. A common problem with these methods is that they either cost money (the cost of sending photos over the mobile network depends on your data communication plan), or the destination computer or your phone may lack a Bluetooth or infrared communication feature.*

tip *Some phone models come with synchronization software that lets you copy information between a phone and a computer. The synchronization software requires a data cable between your phone and computer or a Bluetooth or infrared wireless connection. If you have synchronization software installed on your computer, by all means, try it. Synchronization products often have limitations for what information they can copy; for example, ringtones can't always be copied.*

It's important to realize that phones equipped with a memory card come with two separate memory spaces:

1. *Phone's internal memory*. This is the memory space inside the phone that you can only access with a software utility, like File Manager.
2. *Memory card*. This is a physical medium where you can save information, remove it from the phone, and insert it into another device or computer.

If your photos have been saved on the phone's internal memory, you need to move them to the memory card. In order to make the process easier the next time you want to copy photos to your computer, change the camera settings on the phone so that the photos are automatically saved on the memory card. Here's what you have to do:

Symbian OS/S60 Phone Activate the camera, Open the Options menu, and click Settings. Scroll down until you are highlighting Memory in Use. Click to change the value to Memory Card (see Figure 2-1).

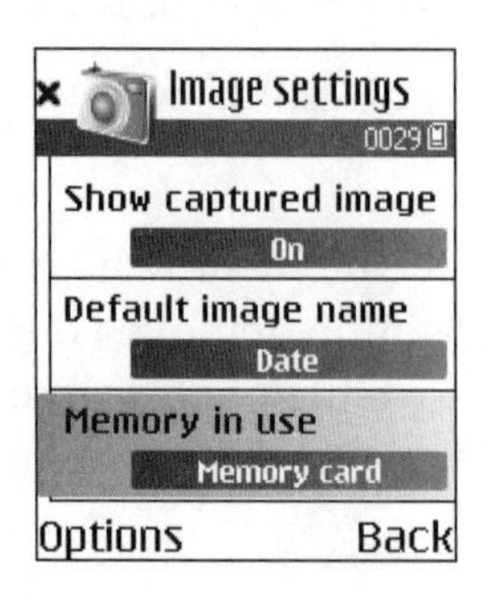

Figure 2-1

Set the memory card as the storage space for your photos on a Symbian OS/S60 phone.

Windows Mobile Phone Activate the camera, open the menu, and select Settings | Storage Folder | My Device | Storage Card (Figure 2-2). When you open the menu, but have Options entry instead of Settings, select Options | General | Storage | Storage Card.

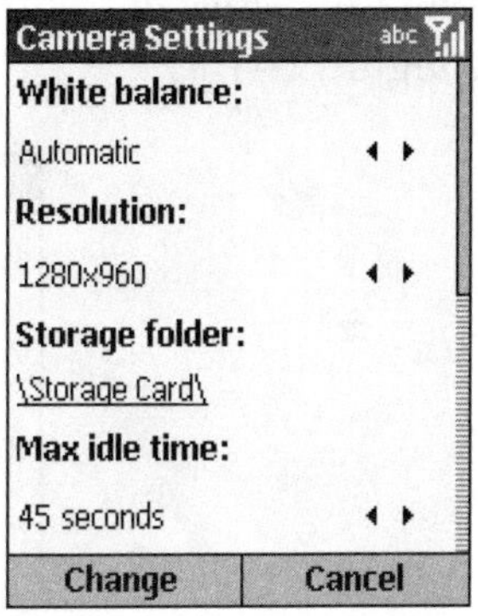

Figure 2-2
Set the memory card as the storage space for your photos on a Windows Mobile phone.

Other Phones Activate the camera, open the camera menu, and look for an option that lets you set the memory card as the storage space for photos.

You are going to copy information from the phone's internal memory to the memory card. A word of warning before you start: be careful that you don't touch any files in the phone memory or on the memory card that you don't recognize as your own.

Step 1: Move Your Photos to the Memory Card (If They Are Not Already on the Card)

If your photos have been stored in the phone's internal memory, you have to move them to the memory card. If the pictures are already on the memory card, you can skip to Step 2.

Symbian OS/S60 Phone Go to the main menu. Open the Gallery application (you can find it in the Media folder or another folder) and choose Images. Open the Options menu, choose Mark/Unmark | Mark All. Reopen Options, scroll down until you find the option Organize, and select Move to Memory Card (see Figure 2-3).

Figure 2-3
Move your photos to the memory card on a Symbian OS/S60.

Windows Mobile Phone Go to the Home screen. Hit the Start key and look for the Photo Album or Photos & Videos application. Launch the application when you find it. Locate the photo you want to move (you don't have to open it). Open the menu and select Move. All folders on the phone are displayed. Scroll down until you find Storage Card and select OK.

If you don't find Move entry in the menu, use the File Manager software to move pictures (see Figure 2-4).

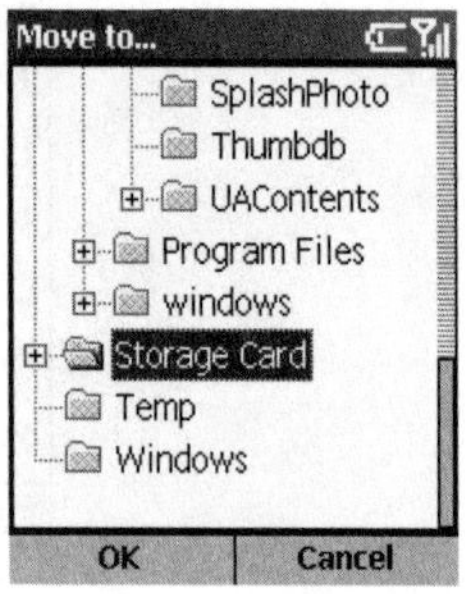

Figure 2-4

Move your photos to the memory card on a Windows Mobile phone.

Go to the Home screen, push the Start key, and launch the File Manager application. Navigate to the folder where the camera has saved your images. My Documents\My Pictures is often the default folder for photos. Once you have discovered your pictures, open the Menu. Choose Selection | Select Multiple. You can choose the photos you want to move by ticking the box in front of each item. Push the selection key for ticking a box. Hit the Done key when you are ready. Open the Menu and select File | Move to. Move the pointer to highlight the Storage Card. Push the Done key.

Other Phones Open the photo album application on the phone and look for an option that lets you move photos to the memory card.

Step 2: Remove the Memory Card from the Phone

Most new phones have a memory card slot on the side, top, or bottom panel of the unit. Often, a cover protects the slot. When you want to remove the memory card from your phone, open the cover and lightly push the memory card until you hear a click. The card should eject so that you can pull it out (see Figure 2-5).

If you can't locate the memory card slot at all, but you know there should be one, remove the battery and peek under it. Some manufacturers still like to hide the memory card in this awkward location.

Step 3: Plug Your Memory Card Reader into the Computer's USB Port

Now, let's move over to your computer. Connect the USB cable from your memory card reader into the USB port on the computer (see Figure 2-6). If you have a computer with a built-in memory card reader, use that.

Figure 2-5

Push the memory card lightly. It will eject and you can pull it out.

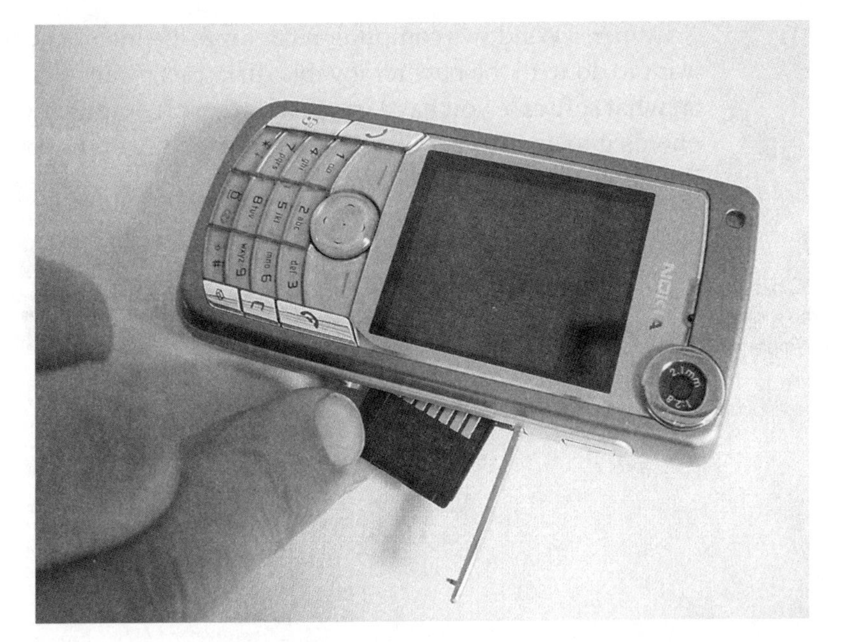

Figure 2-6

Plug your memory card reader into the flat and wide USB connector.

Step 4: Insert Your Memory Card into the Reader

Memory card readers have slots for different card sizes. You should stick the card in a slot where it comfortably fits. If you have successfully inserted the card into the reader, your computer will display a window asking what you want to do with the storage medium. If nothing happens on the computer in a few seconds, check that the card is pushed in all the way to the end. Another trick is to remove the card, turn it upside down, and retry.

When a Windows computer recognizes the memory card, it will ask you what you want to do with the new removable disk (this means your memory card). Depending on what software you have installed on your computer, you can choose to import the photos into an image editing software, a photo album, or another application. Select Windows Explorer (see Figure 2-7).

Figure 2-7

When your PC has recognized the memory card, select Windows Explorer in this window.

Step 5: Copy Photos from the Memory Card to Your Computer's Hard Disk

The Windows Explorer lists the folders on the memory card. First, click Folders at the top of the window. This allows you to manage several storage devices in one window. In the left frame, locate the destination folder where you want to save the photos.

In the window on the right, you should still see the folders on the memory card. Find the photos, grab them with your mouse, drag them to the destination folder on your hard disk, and drop them in there—preferably to a folder reserved for photos (see Figure 2-8).

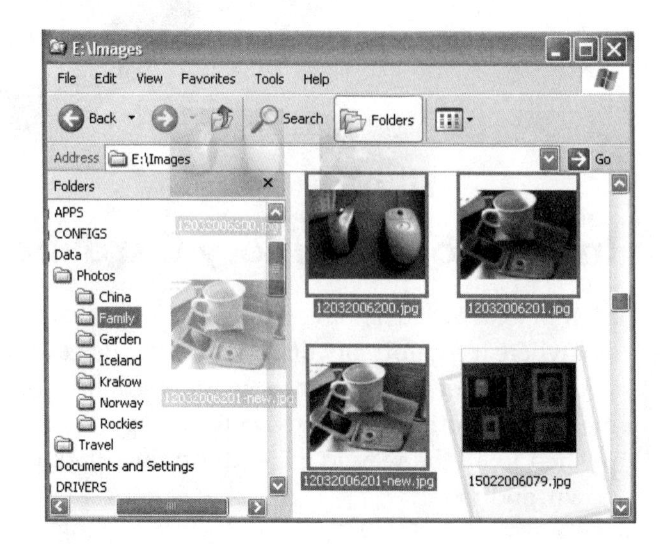

Figure 2-8

Grab the pictures with your mouse and drag them to the folder where you want to store them.

Photos are now saved on your computer's hard disk. You can e-mail your pictures, post them on online photo albums, and archive them on CDs. After you have made a backup copy of photos on a CD, or on any other media, you can delete the photos from the phone's memory card.

Step 6: Make Space for New Photos on the Memory Card

You can choose from two techniques for deleting old photos from your memory card: (1) keep the memory card in the card reader and use your computer to remove old photos, or (2) insert the memory card into your phone and use the phone menu system to remove photos.

While you still have Windows Explorer running on your computer and the memory card is still in the reader, it is easy and quick to delete any items from the memory card. Simply select the photos you want to remove and press DEL on your computer keyboard. If you want to delete several photos, hold down CTRL while you click with the mouse each photo to be deleted. If you hold down SHIFT and click on the first and last item, a range of items will be selected.

Project 3

Copy Photos and Other Information from Your Phone to a PC Using a Wireless Bluetooth Connection

What You'll Need:

- **Bluetooth connectivity on your PC. If your computer lacks Bluetooth, you can add easily add it by purchasing a small USB adapter**
- **Bluetooth connectivity on your phone. If your phone didn't come with Bluetooth, you can't add it afterwards**
- **Cost: $20–$30 U.S. for the Bluetooth adapter**

When you want to move photos, songs, or other information from your phone to a PC, you can do it without any cables or memory cards. Bluetooth wireless technology has been designed to replace cables between phones, other mobile devices, computers, and peripherals. Bluetooth is used for transmitting information such as e-mails, documents, and photos from phone to phone, or from phone to computer. Bluetooth can also connect wireless headsets to phones.

Typically, Bluetooth products transmit data at the speed of 720 kilobits per second (Kbps), but new products can achieve higher speeds as well. The signal range is about 30 feet (10 meters), but there are products available with longer range. Bluetooth signals are transmitted over radio waves and don't require a line of sight between communicating devices. Bluetooth works both ways: you can also copy photos and other information from your computer to your phone.

Consider using Bluetooth instead of a memory card for moving information, when

- You only want to move a small amount of information between your phone and PC. For example, it is convenient to transfer a few photos via Bluetooth, but transferring MP3 songs is rather slow and may test your patience.
- You are using a dedicated application, such as synchronization software that relies on Bluetooth for connectivity.
- There is no memory card on your phone.

note *Infrared is another wireless technology for transferring information between devices. For example, phone-to-phone and phone-to-PC connections are common usages for infrared. Although an established technology, infrared is being replaced by Bluetooth in modern computers and phones. Infrared transmission requires a line of sight between the devices, which makes it sensitive to transfer errors. Although this project shows you how to use Bluetooth, you can apply these instructions to infrared connectivity as well. You just have to be ready to explore the options for infrared on your PC and on your phone.*

tip *If you are not sure if your PC is equipped with Bluetooth, you can easily verify it. Open the Control Panel application on your PC. Look for an entry labeled Bluetooth (it may be labeled Bluetooth Configuration, or Bluetooth Com, but as long as it says Bluetooth, you're fine). If you can find it, open it. You should be able to view a list of COM-ports for Bluetooth. If you can't, you may purchase a low-cost adapter and install it into a free USB port in your computer (see Figure 3-1). Follow the installation instructions included with the product.*

Figure 3-1
Bluetooth adapter installed in a USB port.

Step 1: Activate Bluetooth on Your PC

First, make sure Bluetooth is switched on in your PC. Go to the Start menu and launch Windows Explorer. Click Folders at the top of the Windows Explorer window. Scroll down the left frame until you find Bluetooth. The title may be Bluetooth Information Exchanger, Bluetooth Neighborhood, or something similar, but as long as it says Bluetooth, that's the one you are looking for (see Figure 3-2). Click the Bluetooth entry in the left frame.

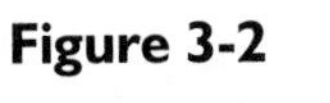

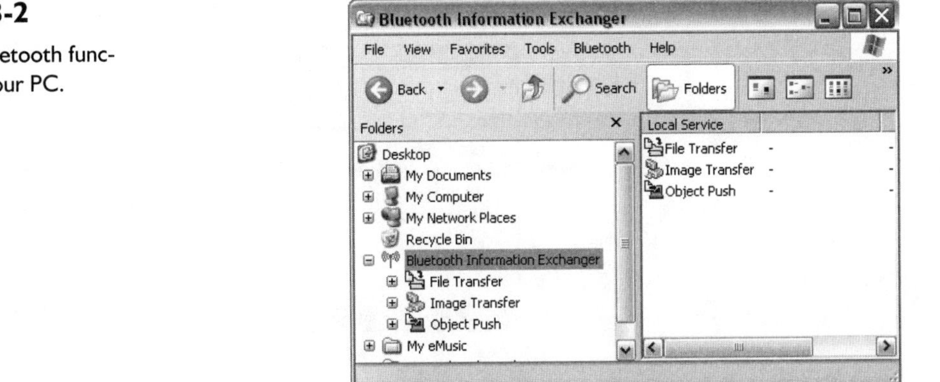

Figure 3-2
Access Bluetooth functions on your PC.

The right frame of Windows Explorer should display folders and information about Bluetooth connectivity. If nothing happens, or the listed items are inactive, you must verify that the adapter is connected and switched on.

Step 2: Activate Bluetooth on Your Phone

Many phone models display the Bluetooth symbol on the home screen if the connection is active. You can skip to step 3 if Bluetooth is already switched on in your phone.

Symbian OS/S60 phone Go to the main menu and open the Connect folder. Find an entry titled Bluetooth and open it. You'll see the current settings for Bluetooth. If the value for Bluetooth is Off, highlight it, push the selection key, and set the value to On (see Figure 3-3).

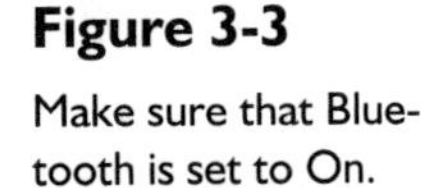

Figure 3-3
Make sure that Bluetooth is set to On.

While you are viewing the setup screen, you might as well change the Bluetooth name for your phone. The name can be anything you want, but keep in mind that this name will be viewed by people who want to send you something from their Bluetooth-equipped phones and computers. Select My Phone's Name and type in the name you want.

Windows Mobile Phone Go to the Home screen and push START. Select Settings. Look for Bluetooth in the list of items (note that the list may be several pages long). Open it and change the Bluetooth mode to On or Discoverable (see Figure 3-4). Push DONE.

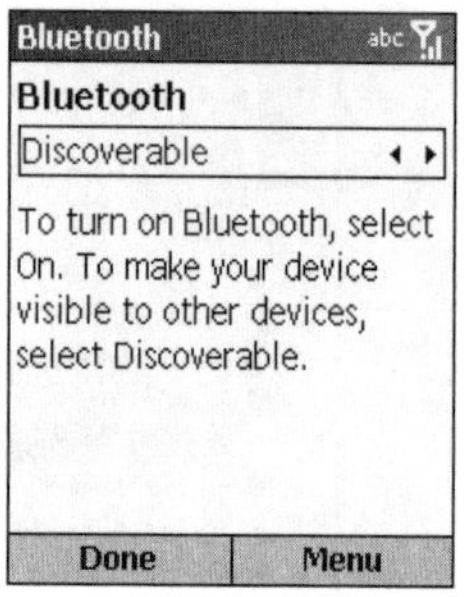

Figure 3-4

Turn Bluetooth to Discoverable or On when you want to use Bluetooth.

If you couldn't find Bluetooth entry in the Settings menu, you should be able to see Connections in the same menu. Open Connections. Select Bluetooth and change the value to Discoverable or On. Push DONE.

On Windows Mobile devices, the Bluetooth mode On means you can transmit information, but other devices won't see the name of your phone when they are scanning for Bluetooth devices. Discoverable means that your phone can be detected when other devices are scanning for Bluetooth connections.

Other Phones Find Bluetooth from the phone menus and activate it.

Next time, when you want to connect your phone and PC over Bluetooth, you don't have to go through steps 1 and 2 anymore if you leave Bluetooth on. If you know that you won't need Bluetooth for a while and want to be sure that no one is able to connect to your phone via Bluetooth, you can turn it off.

To avoid problems from harmful software and hacking attempts, don't accept connections from devices that you don't know.

Step 3: Send the File

Now, you are ready to send a file from your phone to your PC. Let's use a photograph as an example file. Some phone models restrict sending of items like ringtones over Bluetooth, but you can always send photos that you have snapped yourself.

Symbian OS/S60 Phone Go to the main menu. Open the Gallery application that is usually located in the Media folder, but check other folders if you can't find it in there. Select Images and find the picture you want to send. Open the Options menu and select Send | Via Bluetooth (see Figure 3-5).

Figure 3-5
Send a picture via Bluetooth.

The phone searches for Bluetooth devices in the range. When the scan has finished, highlight your PC's Bluetooth name and hit SELECT (see Figure 3-6).

Figure 3-6
Select your PC from the discovered Bluetooth devices.

Windows Mobile Phone When you are in the Home screen, push START. Find the Photo Album or Pictures & Videos application and open it. Highlight the photo you want and hit MENU. Select Send via | Bluetooth (see Figure 3-7). If your phone menu includes an entry labeled Beam, select it instead of Send.

Figure 3-7
Send a picture via Bluetooth.

The phone scans for Bluetooth devices in the range. Wait a moment until the scan finishes and highlight your PC's Bluetooth name. Push SELECT (see Figure 3-8) or BEAM (depending on your phone model).

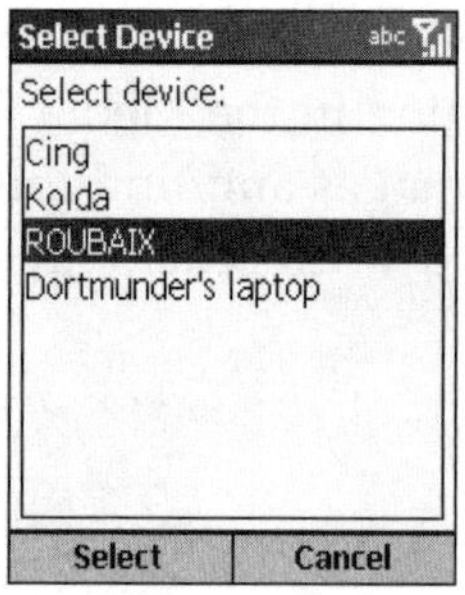

Figure 3-8

Select your PC from the discovered Bluetooth devices.

Other Phones Open an application, such as photo album, that lets you view images. Select a picture, and send it via Bluetooth. When the phone has finished searching for Bluetooth devices, select your PC from the list.

tip *You can send pictures to other phones, printers, and any other devices that have Bluetooth connectivity. When your phone has discovered nearby Bluetooth devices, select the device (instead of your PC) where you want to send your photo.*

Step 4: Save the File on Your PC

When you have sent the file from your phone, keep an eye on your PC. It depends on the manufacturer of your PC's Bluetooth software what exactly you have to do in this step.

note *Some Bluetooth applications receive the file and save it in a default folder without asking for confirmation. You have to find the received file from the hard disk yourself. A good place to start is Windows Explorer. Open Windows Explorer, click Folders at the top of the window, and choose Bluetooth in the left frame. Click the folders under the Bluetooth folder and view their contents for the received file. Another way to find the received file is to right-click the Bluetooth title in the left frame of Windows Explorer. Choose Properties from the pop-up menu, and try to find the name of the folder where the Bluetooth software saves the received files.*

Usually, Bluetooth software asks for your confirmation before it receives or saves any files sent over the wireless connection (see Figure 3-9). In this case, you can select the folder where you want to save the file.

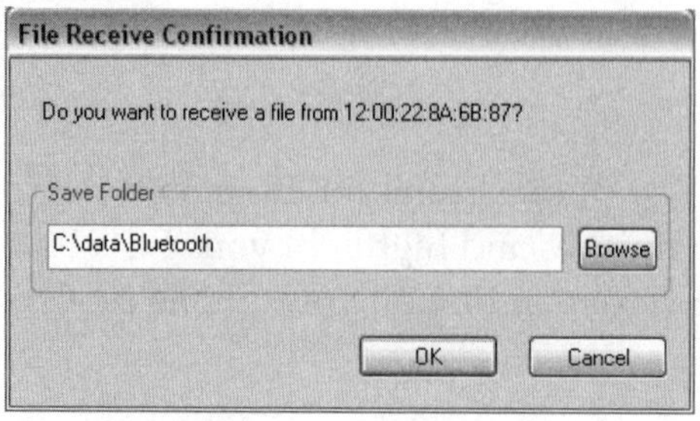

Figure 3-9

Save the received file on the hard disk.

note *Some Bluetooth devices require you to confirm the transfer both on the sending and receiving devices. In this case, the receiving device will ask for a Passcode or Passkey when you initiate the transfer. You can type whatever you want into the Passcode field. The point is to enter the same code in both devices so that you know that you are accepting files from a source you trust. For example, type 123 and hit OK. Next, the sending device will ask for the Passcode. Type the same Passcode (123) into the sending device, and the transfer can begin.*

Project 4

Record Home Movies with Your Camera Phone

What You'll Need:

- **A memory card for storing your videos**
- **Cost: $10 U.S. or more, depending on the card's storage capacity**

Practically every camera phone is a digital video camera as well. Although the videos recorded with a camera phone may not turn out exactly like Cinemascope movies, shooting the movies, showing them to friends, and posting them to video-sharing sites on the Internet is a lot of fun.

Step 1: Activate the Video Camera

You can find the video recorder when you activate the digital camera on the phone and change the recording mode from Camera to Video. Some camera phones come with a dedicated application for video recording. In this case, you have to launch the video recorder from the menu system.

Symbian OS/S60 Phone Activate the camera by pushing a dedicated camera key, or launching it from the phone menu. Open Options and select Video mode (see Figure 4-1).

Windows Mobile Phone Activate the camera by pressing a dedicated camera key, or opening it from the menu. Open the camera menu, and select Record Video (see Figure 4-2). If you can't find Video in the menu, select Capture Mode | Video.

Other Phones Activate the camera, open the menu, and change the recording mode to Video.

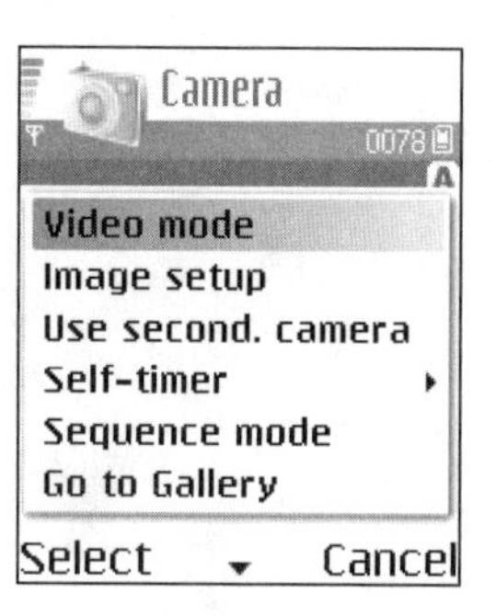

Figure 4-1

Set the camera on your Symbian OS/S60 phone ready for video recording.

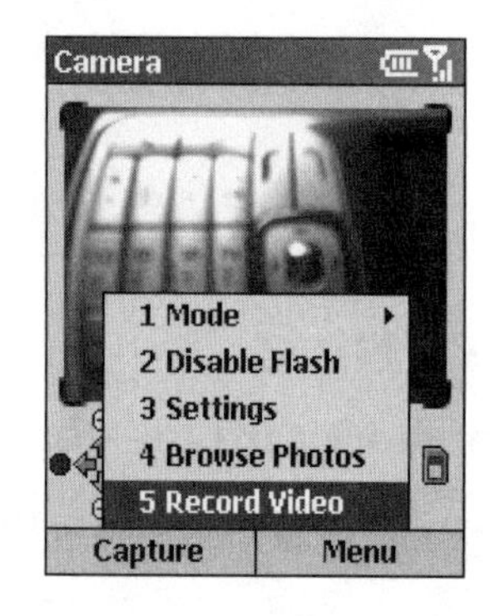

Figure 4-2

Set the camera on your Windows Mobile phone ready for video recording.

note *When you start shooting a video, several things happen in the phone. The image sensor captures multiple images per second, and the imaging software on the phone compresses the images. The compression usually removes information that doesn't change from frame to frame; for example, the blue sky in the background doesn't have to be stored in every frame. You'll notice the compression because video images are not as sharp as still images and occasional mosaic-like image sequences may show up as well. When you stop recording, the clip is saved on the memory card or in the phone memory (depending on your camera settings).*

Step 2: Set the Default Storage

If you haven't already done so, make sure that the recorded videos will be stored on a memory card instead of in the phone's internal memory.

Symbian OS/S60 Phone When the video camera is active, select Options | Settings | Memory in Use | Memory card.

Windows Mobile Phone After you have activated the camera, open the Menu. Select Options | General | Storage. Pick up Storage Card from the list.

Other Phones Activate the camera, open the Menu, and change the default storage to memory card.

Step 3: Set the Video Size

Your camera phone may allow video recording in different resolutions. You get the best quality when you set the video size to the largest you can find. This may increase the size of the resulting file considerably. If you intend to send the clip as an MMS (Multimedia Message Service) message, you should keep the video file size under 100KB. That's the maximum MMS message size that some networks let through. You can, however, always send large attachments by e-mail (and you don't have to pay extra for the MMS service).

Symbian OS/S60 Phone When the video camera is active, select Options | Settings | Video Quality | High (see Figure 4-3).

Figure 4-3

Set the highest video resolution if you intend to view your home movie on a computer screen.

Windows Mobile Phone When you have activated the video camera, open the Menu, select Options | Record Quality | High (see Figure 4-4). If you don't see the Record Quality choice in the Options, select Modes | Resolution, and choose the highest resolution you can find. You may also change the Encoder value to AVI, or MPEG-4 for better image quality.

Figure 4-4

In addition to the video quality, you can choose to record audio as well.

Other Phones Open the camera menu and set the highest resolution you can find.

tip *Many camera phones capture movies at the speed of 8–10 frames per second. The common picture size for video captured by camera phones is 176×144 pixels. This is also known as quarter common intermediate format (QCIF), which originates from videoconferencing standards. This video size looks fine on a phone screen, but it is very small if you view it on a computer screen. Some of the latest two- and three-megapixel camera phones can record video in 320×240 pixel resolution, at 15 frames per second. Movies captured with a camera phone like that are a pleasure to watch even on a computer monitor.*

Step 4: Begin Recording

With all the required settings in place, you can start recording your first short film.

Begin recording by pushing RECORD, or the selection key you would use for confirming actions, such as the center key on a four-way navigation pad.

Step 5: Keep the Phone as Steady as You Can

If possible, let the scene or objects in front of the camera move without moving the camera. The idea is to keep the lens steady, because even the slightest movement will be visible on the video.

Step 6: When You Move the Lens, Move It Slowly and Steadily

If you don't move the lens slowly and steadily, you won't see what's happening on the screen and the viewers of the video will only see very blurry or mosaic images.

Step 7: Don't Use Digital Zoom

There's no optical zoom in a typical video recorder on a camera phone. You should use your feet to get closer to the subject of your movie, instead of using the digital zoom feature that degrades image quality.

Step 8: Recording Movies Requires Plenty of Light

There's rarely enough light in a normal household for shooting camera phone videos indoors. Unless the indoors lighting conditions are exceptionally good, go outside and record your home movies in the daylight.

Step 9: Record Short Clips Rather Than Long Scenes

Some camera phones restrict the length of a single take to 10–30 seconds, and often, that's a good goal. If you can't film your Oscar winner in 30 seconds, you may not have any idea what you are shooting in the first place. Most camera phones, however, let you record as long as there is available memory space for storing the video.

Stop the recording by hitting STOP. Now, you have to be careful. Most phones automatically save the clip once you press STOP, but some phones require you to specifically select Save to store the video in the memory.

note *With digital still photos, you don't have to worry about the file format of the images—they are practically always in JPG format. There's more choice in digital video as the high-tech world hasn't yet come to an agreement regarding what should be the universal format for digital video. Luckily, the situation with camera phone video formats is somewhat better. Third Generation Partnership Project has defined a standardized video format, known as 3GP. Most camera phones can record and play this format. 3GP is used in MMS messages as well. High-end camera phones can also record in Audio Video Interleave (AVI) or Motion Picture Experts Group (MP4, also known as MPEG-4) video formats. These formats produce larger video files, but of much better quality than the 3GP (see Figure 4-5).*

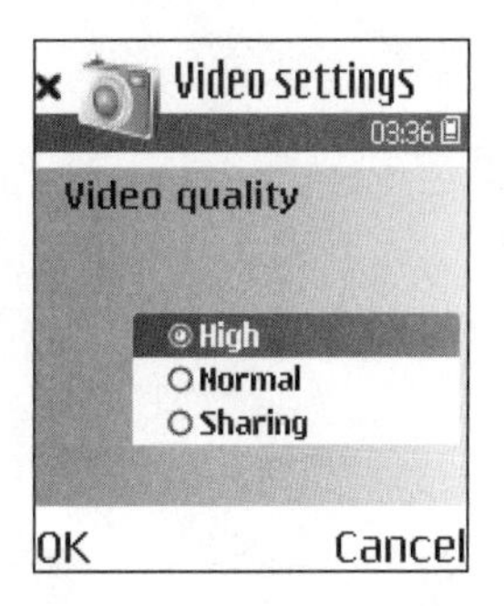

Figure 4-5

A two-megapixel camera phone that can record high-quality MP4 video in 320×240 resolution. Normal and low-quality videos are saved in 3GP format in 176×144 resolution.

tip *When you want to watch your new home video on a computer, you need software that can play the video. Popular and free video player software products are Apple QuickTime, RealPlayer, and Windows Media Player. If the Windows Media Player can't play your 3GP video, try RealPlayer.*

tip *The real fun starts after you have copied your video to a computer and can share it with friends or perhaps with the worldwide audience on the Internet. Popular video-sharing services on the Internet are, for example, YouTube (www.youtube.com), Buzznet (www.buzznet.com), Yahoo Videos (video.yahoo.com), and Google Videos (video.google.com). Some of these services can receive video clips submitted directly from the phone as well. Or how about arranging a premier night home movie show in your living room on your large-screen TV for your guests?*

Project 5

Listen to MP3 Music on Your Phone

What You'll Need:

- **A phone with MP3 music player software: Even if your phone came without an MP3 player, you may be able to download a piece of software that turns your phone into a music player. For example Handango (www.handango.com), My-Symbian.com (www.my-symbian.com), and Smartphone.net (www.smartphone.net) offer plenty of MP3 software products**
- **A stereo headset for your phone: A headset designed for the phone allows you to use it both for phone calls and for listening to music. Music phones typically come with stereo headsets. If you intend to buy a stereo headset, make sure your phone is capable of stereo sound**
- **A memory card: Any memory card that's compatible with your phone and your memory card reader will do. The higher the storage capacity is, the more music you can have. For example, a 512 MB card can store up to 150 tracks of high-quality MP3 music**
- **A memory card reader: Use a reader that you can attach to a USB port on your computer, or use a built-in memory card reader on a notebook computer**
- **Music library: You need an application, like iTunes, Musicmatch, or Windows Media Player that can play MP3 music, extract songs from CDs, and organize tracks into a collection**
- **Cost: A 512 MB memory card costs about $20–$35 U.S. You can get a memory card reader for $15–$30 U.S. Smartphone owners can purchase an MP3 application for $15–$25 U.S**

Have you realized that you may already own a portable MP3 music player that you are always carrying with you? That's your phone. By creating a digital music library on your computer and preparing your phone for MP3 music, you can take your favorite music wherever you go, without carrying extra weight.

In many ways, a phone can be a more convenient music device than a dedicated MP3 player, such as the Apple iPod or the Creative ZEN. For example, when you are listening to music through a headset and the phone rings, the music automatically pauses. You can hear the phone ring and you can see who's calling, even when the music is playing. After the call, the music resumes where it was paused. If you receive a message to your phone in the middle of a long podcast, you can read it while listening to the podcast.

When you are ready to turn your phone into an MP3 player, begin by establishing a music library on your computer. Then, save MP3 songs from your CDs and other sources into the library. Finally, identify the technique that best suits you for transferring music to your phone.

Step 1: Create a Music Library on Your Computer

Even if you are thinking that you can quickly download your favorite CD to the phone, and won't ever need any more music on the go, take the time and establish a music library on your computer. The music library will become the central storage for all your digital music.

Popular products for managing digital music libraries are, for example, iTunes (see Figure 5-1), Musicmatch, and Windows Media Player (see Figure 5-2). They let you extract tracks from CDs, download songs from online shops, and sync music with portable MP3 players. Each of these applications is available as a free download.

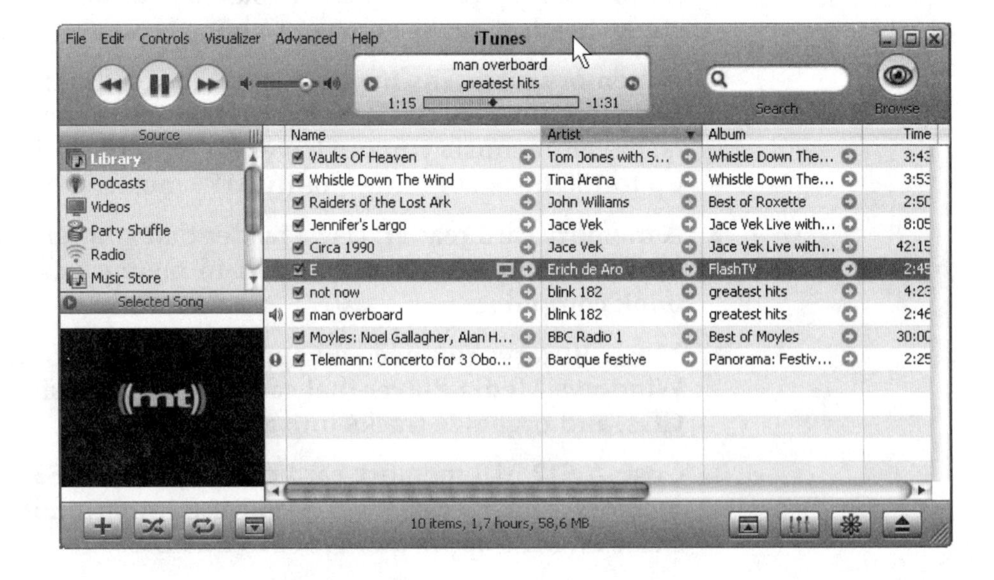

Figure 5-1

The main screen of iTunes 6 shows your the songs stored in the music library. iTunes 7, which looks similar to iTunes 6, includes the possibility to purchase movies in addition to music.

Figure 5-2

The music library in Windows Media Player 11. You can view your library by album, artist, song, or other criteria.

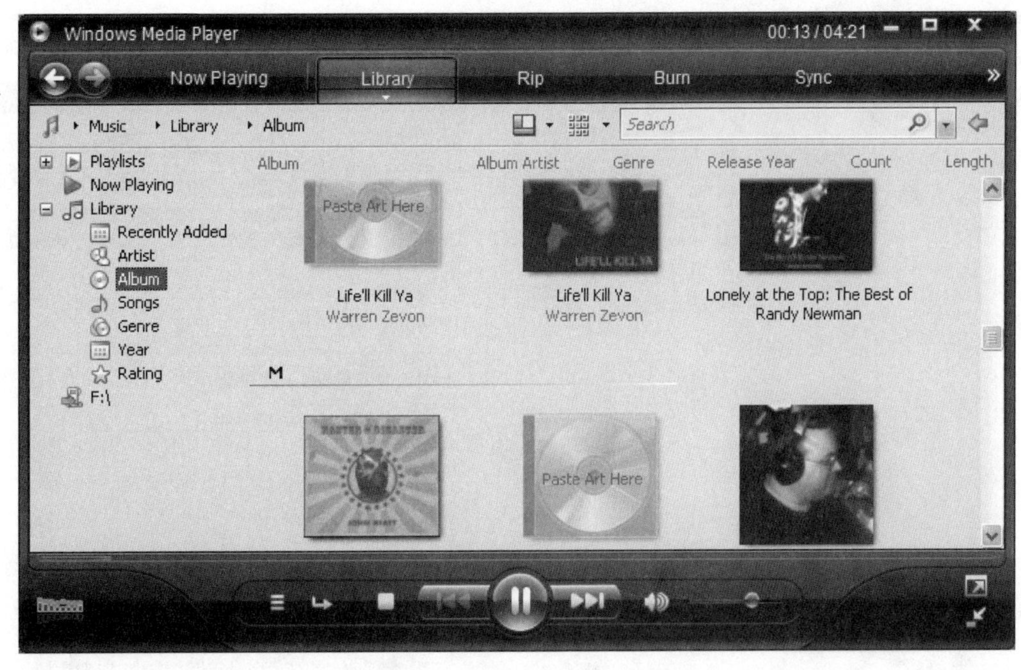

Step 2: Turn Your Music into MP3 Songs

When you think about it, there is music all around you: on CDs, on the radio, on old vinyl records, and on the Internet. Make use of what you have, and save the music you already own on your computer.

If you have a collection of audio CDs, start with them. Use iTunes, Musicmatch, Windows Media Player, or another software to extract (some products call it ripping, or importing) tracks from the CDs onto your computer's hard disk. The important thing is to extract the tracks as MP3 files (see Figure 5-3). Otherwise, your phone may not be able to play the songs at all.

You can find more details on copying tracks from music CDs to a computer in Project 9.

note *It takes only a few clicks to download tracks from online stores. Popular music download stores are, for example, iTunes, Napster, and Rhapsody. They are popular, but here's the really tricky thing: you shouldn't be shopping in these stores for music for your phone. These and many other online shops sell music that is copy-protected. It is likely that copy-protected tracks won't play on your phone. There are a few exceptions, like Motorola's iTunes-compatible phone, or some Nokia and Samsung phones that are compatible with copy-protected Windows Media -encoded tracks. Nonetheless, as a rule of thumb, don't purchase any copy-protected music.*

There are shops that offer high-quality MP3 music that isn't copy-protected. Examples of online stores that provide MP3 tracks that you can play on any of your digital music devices, including your music phone, are eMusic (www.emusic.com) (see Figure 5-4), eClassical (www.eclassical.com), and Wippit (www.wippit.com).

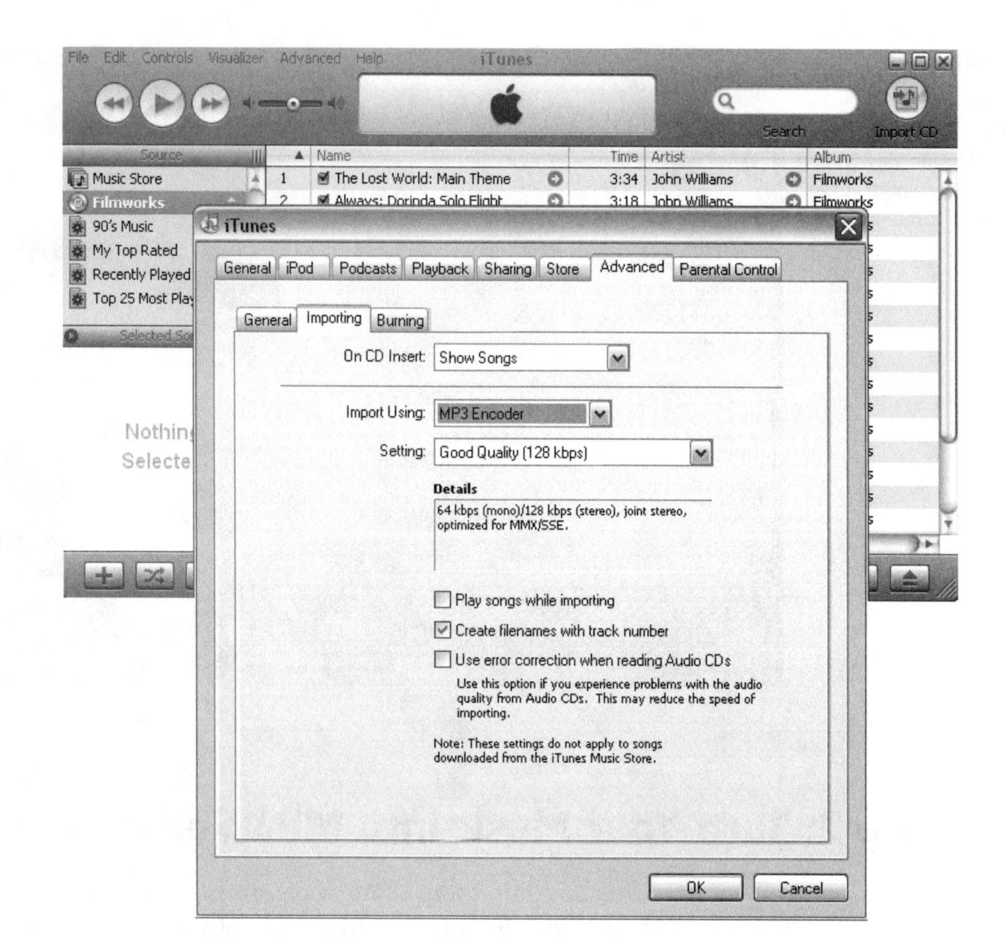

Figure 5-3

Set the music format to MP3 when you import songs from CDs.

tip *A (relatively) long time ago music lovers had to buy huge black vinyl discs when they wanted to listen to their favorite music. If you have a collection of those scratchy treasures, you can save songs from them on your computer. All you need is software that can record from the computer's line-in port. The Sound Recorder application included with Windows XP, RealPlayer or one of the many recording applications available at Download.com can help you save your precious vinyl LPs. Connect your old gramophone to the line-in port on your computer and record the old favorites as MP3 files onto the hard disk.*

It is possible to avoid the use of a computer altogether and download music directly to the phone. Many service providers offer songs to their subscribers for download. However, tracks purchased from service providers' online shops typically have restrictions on which devices they can be played. For example, you may not be able to play the purchased music on another phone or on a dedicated MP3 player.

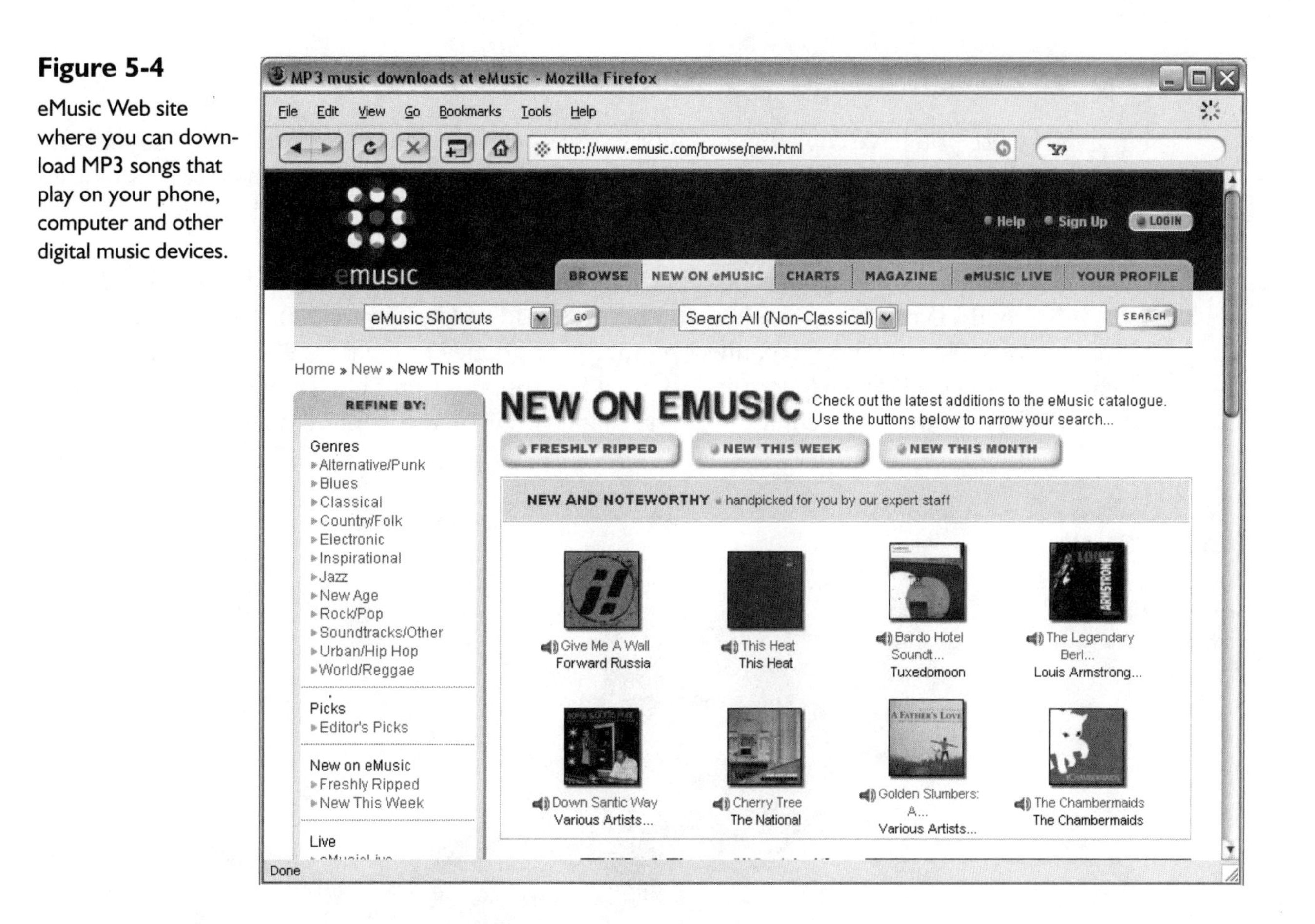

Figure 5-4

eMusic Web site where you can download MP3 songs that play on your phone, computer and other digital music devices.

Step 3: Copy Songs from Your Music Library to the Phone

Transferring music from your computer to the phone is a snap after you've done it once. Dedicated MP3 players often come with special applications, like iTunes, that can automatically synchronize the portable device with new music saved on a computer. A few phone models, for example, from Motorola and Nokia, can also do this. However, without any of those products, you can still copy music from your music library to a phone that can play MP3 tracks.

The most flexible way to copy and store a number of songs on the phone is to use a memory card. Use the File Manager software on the PC to drag and drop MP3 files from your music library to the phone. Read project 6 for details about using a memory card for transferring information to the phone.

A wireless Bluetooth connection is another useful method for quickly copying a song or two from your computer to the phone. Project 7 outlines the steps for achieving this.

A data cable is a reliable tool for transferring large amounts of information between a phone and a computer. If your phone included a data cable and synchronization software, check it out. Some products let you copy songs from your computer to the phone.

Step 4: Listen to MP3 Music on the Phone

Once you have copied your favorite tracks to the phone, open the Music Player (Media Player or MP3 player) application on your mobile device. Some players require you to update the tracklist before they can play the recently added songs.

All phones have loudspeakers, and music phones typically come with stereo speakers. Although it is possible to listen to music through the loudspeakers only, the sound is much better when you use the headset (see Figure 5-5). A number of music phones allow you to connect regular headphones to the phone's headset connector. If you do that, keep in mind that the headphones don't have a microphone and you have to unplug them if the phone rings while you are listening to music.

Figure 5-5

A phone that is also an MP3 music player with stereo sound.

Project 6

Copy Ringtones, Wallpapers, MP3 Music, and Other Fun Stuff to Your Phone Using a Memory Card

What You'll Need:

- **A memory card that is compatible with your phone**
- **A memory card reader that you can attach to a USB port in your computer. Alternatively, a computer with a built-in memory card reader can be used as well**
- **Cost: $10–$50 U.S. for a memory card, depending on its storage capacity. $15–$30 U.S. for a memory card reader**

You can copy practically any item, such as ringtones, MP3 songs, podcasts, photos, or useful software applications from your computer to your phone. This project will show you how you can use a removable memory card for copying these items.

- MIDI tunes Needed if you want to download ringtones to your phone. Other ringtone formats are widely used as well, but MIDI tunes are easy to find. MIDI music is compatible with most phones that can play polyphonic ringtones. WAV audio is another universal format for ringtones, but it requires more memory space then MIDI. Advanced phones can play MP3 tracks as ringtones as well.

- MP3 files Needed if you want to copy music or podcasts to your phone. Any phone that features a music player should play MP3 tracks. If you know what you are doing (for instance, if you own a phone compatible with iTunes), you might try some other music formats as well. Unless you have special requirements for your digital music, use only MP3 since practically any digital music device can play MP3.
- Videos in 3GP format Needed if you want to watch movies or video clips on the phone. Phones that come with Windows Media Player or RealPlayer software are able to play other video formats as well. Still, 3GP is the only video format that is recognized by many different phone models. If the video you want to transfer to your phone is not in 3GP format, you can convert it into 3GP using video conversion software.
- JPG pictures Needed if you want to have new wallpapers or photos on the phone.
- Software applications (compatible with your phone) You can install SIS applications to Symbian OS phones, CAB applications to Windows Mobile phones, and Java (JAR and JAD) applications to most modern phones.

There are other methods that let you transfer information from your PC to the phone, such as Bluetooth, infrared, or data cable, but they often have some limitations. The benefit of the memory card technique is that no software, hardware, network, or service provider can restrict what you copy. Also, the procedure is principally the same regardless of the phone model.

note *It is possible to copy (1) music you have extracted from your CDs, (2) podcasts and videos you have downloaded from the Internet, and (3) photos you have snapped with your digital camera. As long as the material is stored on your computer's hard disk, it can be transferred to a memory card. However, if you have copy-protected items, such as songs purchased from the iTunes Music Store or Napster, they'll remain copy-protected (and probably won't play on your phone), no matter where you move them.*

Step 1: Plug a Memory Card Reader into a USB Port on Your Computer

Connect the USB cable from your memory card reader to the computer (see Figure 6-1). If there's a built-in memory card reader on your computer, you may be able to use it instead of a dedicated card reader.

Figure 6-1

Connect a memory card reader to your computer.

Step 2: Remove the Memory Card from Your Phone

Memory card slots are usually located on the side, bottom, or top of the phone. If you can't find your memory card, open the battery cover, remove the battery, and take a look under the battery (see Figure 6-2). Memory cards have been caught hiding in battery compartments as well.

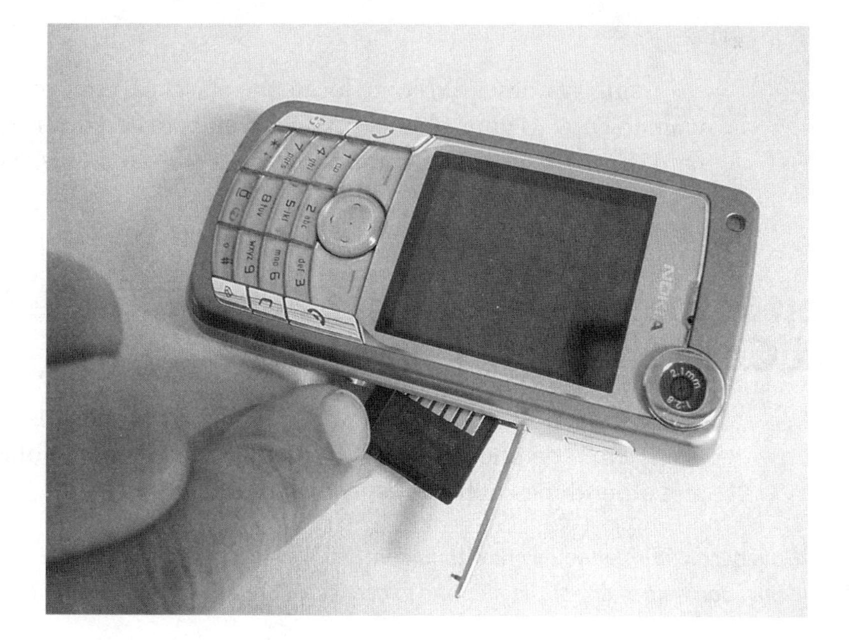

Figure 6-2

Open the card slot cover, lightly push the memory card, and pull it out.

Step 3: Insert the Memory Card into the Card Reader

After you have inserted the memory card into the reader, wait for a few seconds for a window to pop up on your PC screen. If nothing happens, remove the memory card, turn it upside down, and stick it back into the reader.

Step 4: Select Windows Explorer in the Pop-up Window

When the pop-up window is asking what you want to do with the new disk drive (it means your memory card), scroll down and select Windows Explorer (see Figure 6-3).

Figure 6-3
Select Windows Explorer for viewing the contents of your memory card.

Using Windows Explorer, locate the object on your computer's hard disk that you want to copy to your phone. If you haven't created your own folder structure on the hard disk, see what you can find in the My Documents, My Music, My Videos, and My Pictures folders.

Step 5: Identify the Target Folder on the Memory Card and Copy the New Items to the Card

The key to success in copying new items to the phone is finding the right folder for each item. The phone expects to find ringtones in one folder, wallpapers in another, and other things in their respective folders.

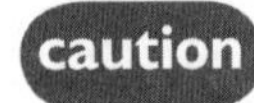

Be very careful when you work with folders and files. Unless you are absolutely sure what you are doing, don't touch any of the files and folders configured at the factory, or created by applications.

note *Different phone models save information in different folders. This is true even for phones that are powered by the same operating system software. In practice, this means that even though the following instructions have been tested with several phone models, you may have to do some exploring by yourself and apply them to your phone.*

The essential tool in this project is Windows Explorer. Open the application and click on the Folders tab at the top of the Windows Explorer window to view the folders on the memory card. The folders are listed in the left frame.

A) Copy Ringtones to Your Phone

Once you have identified tunes on your computer that may become good ringtones, you are ready to copy them to the phone.

Symbian OS/S60 Phone Copy ringtones from your computer to the Sounds\Digital folder on the memory card (see Figure 6-4).

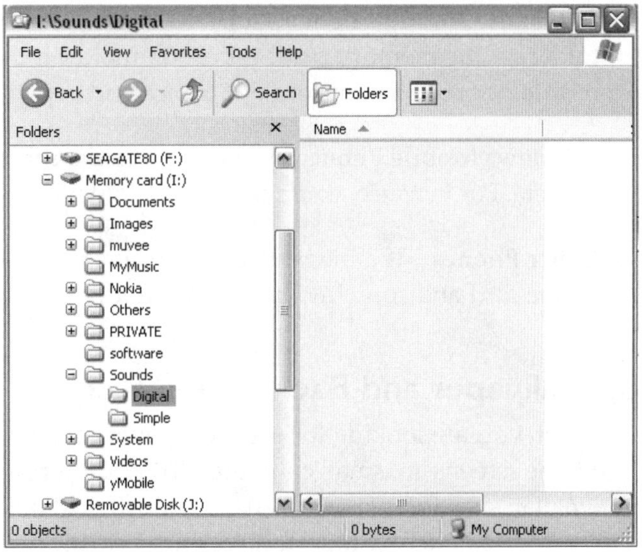

Figure 6-4

Copy ringtones to the Sounds\Digital folder. (This memory card came with a Nokia smartphone and some of the folders were configured at the factory.)

I:\Sounds\Digital
File Edit View Favorites Tools Help
Back Search Folders
Folders
Name
SEAGATE80 (F:)
Memory card (I:)
Documents
Images
muvee
MyMusic
Nokia
Others
PRIVATE
software
Sounds
Digital
Simple
System
Videos
yMobile
Removable Disk (J:)
0 objects
0 bytes
My Computer

Windows Mobile Phone If you have purchased a blank memory card, you can create your own folders on the card. Create a folder for ringtones, another for MP3, one for video, and more for other items you might need. Then, copy your new tune to the ringtones folder (see Figure 6-5).

Other Phones If a memory card was included with the phone, explore the folders on the card and copy ringtones to a folder reserved for them.

B) Copy MP3 Music and Podcasts

Using Windows Explorer, locate MP3 files you have stored on your computer.

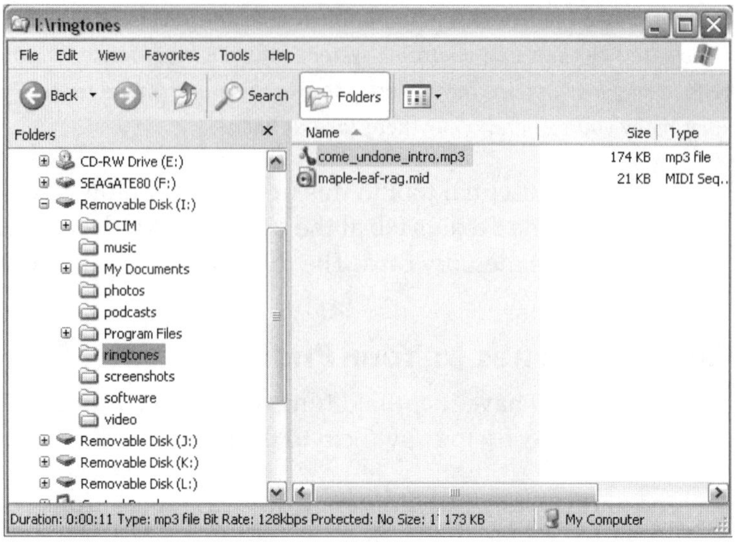

Figure 6-5
Copy ringtones to the memory card. (I purchased this memory card separately from the phone and created folders for music, podcasts, ringtones, and videos.)

Symbian OS/S60 Phone Copy MP3 songs from your computer to the MyMusic folder on the memory card. It doesn't matter if the folder doesn't exist; just create a new folder with a name of your choice and copy your MP3 music in there.

Windows Mobile Phone Make a new folder for MP3 tracks and copy your new tunes in there. For instance, you may create a folder called Music on your memory card.

Other Phones If a memory card was included with the phone, explore the folders on the card and copy MP3 tracks to a folder reserved for them.

C) Copy Wallpaper and Background Pictures

When you are looking for nice wallpaper images for your phone, keep in mind that phone screens are small compared to computer screens. Look for JPG images that are small in size; for instance, 320×240 or 640×480 pixel images make fine wallpapers.

Most phones try to scale large images to the dimensions of the screen, but it doesn't always work. You may have to put faith in your own hands and scale down a large picture using image editing software. If you do that, scale down to 320×240 pixels, or to the native resolution of your phone screen. Common screen sizes in camera phones are 176×220 and 176×208 pixels, but there are many variations.

Symbian OS/S60 Phone Copy the pictures that you want to have as wallpapers to the Images folder on the memory card.

Windows Mobile Phone Copy your new wallpaper images to the top-level folder (designated with the drive letter, for example E:/) on the memory card.

Other Phones If a memory card was included with the phone, explore the folders configured on the card and copy wallpapers to a folder reserved for them.

D) Copy Videos

When you are searching for videos to watch on your phone, keep in mind that practically all camera phones can play videos that are saved in 3GP format.

Symbian OS/S60 Phone Transfer the videos you want to watch on the phone to the Videos folder on the memory card.

Windows Mobile Phone Transfer the video clips you want to watch on the phone to the folder My Documents/My Videos on the memory card.

Other Phones If a memory card was included with the phone, explore the folders configured on the card and copy videos to a folder reserved for them.

E) Copy Software Applications

There are many techniques for installing new software products on your phone. If you purchase an application from an online shop, such as Handango (www.handango.com), My-Symbian (www.my-symbian.com), Smartphone.net (www.smartphone.net), or GetJar (www.getjar.com), the shop usually sends a text message to you. The message includes a link to a page where you can download the product directly to your phone.

Often, free and beta (almost ready to be released) applications have to be downloaded to the computer before you can copy them to your phone. In this case, the installation process is as follows:

- Find the product you want, and download it to your computer.
- Save it on the hard disk and remember the folder where you saved it.
- Copy the software product to the phone's memory card using the Windows Explorer.
- Install the software product using the File Manager utility on the phone. If the software product is delivered in an EXE installation package, unpack it on the computer using Windows Explorer.

tip *Almost all software products designed for phones come with an evaluation period that allows you to try them out for free. The length of the evaluation period is typically between three to 14 days. Once the time is up, the program won't start anymore. If you want to keep the product, you must purchase a license. After the payment, you will get an activation key that unlocks the program and lets you continue with the data and settings you had entered into the phone.*

caution *Before you install any new applications on your phone, make sure that the product is coming from a reputable source—there are some harmful software viruses for phones as well.*

I have created a folder called Software on the memory card that I use as a temporary transfer space for new applications. It doesn't really matter which memory card folder you use, as long as you can access the folder with the File Manager (File Mgr) utility on the phone.

Symbian OS/S60 Phone Copy only applications with the extension .sis, .sisx, .jar, or .jad to a device powered by the Symbian OS operating system. Other types of applications cannot be installed on these phones.

Copy Symbian OS compatible applications from your computer's hard disk to the Software folder on the memory card.

Windows Mobile Phone Copy only applications with the extension .cab, .jad, or .jar to a phone powered by the Windows Mobile operating system. However, not all phone models can install JAD or JAR applications from the memory card; instead these applications may have to be downloaded from the Internet using the phone's browser software.

Copy applications of type CAB from your computer's hard disk to the Software folder on the memory card.

Other Phones Copy JAR and JAD software to the memory card. These are Java applications that run on many modern phones regardless of the brand or model.

Step 6: Remove the Memory Card from the Reader and Insert the Card into Your Phone

When you have copied all the desired items onto the memory card, pull it out from the card reader and insert the card into your phone.

Step 7: Activate the New Items on Your Phone

By now, the new ringtones, wallpapers, and other items should be safely stored on the memory card. There's still one step to go: you must let the phone know about the new items that you want to take into use.

A) Activate Your New Ringtone

Symbian OS/S60 Phone Go to the main menu and open Tools | Profiles. You'll see a list of profiles already defined for the phone. A profile is a collection of sounds that the phone uses as alert tunes when, for example, the phone rings or receives a new text message. For instance, when you are walking along busy city streets, but you are expecting a phone call, you could have an Outdoor profile where all alert sounds ring at the maximum volume.

While you are still viewing the list of profiles, highlight, for example, the Outdoor profile. Open the Options menu and choose Personalise. Select the Ringing Tone field, and you'll be able to view the available ringtones. The list includes the ringtones already on the phone and the tunes you saved on the memory card. The tunes on the memory card have small icons after their titles. Select the ringtone you want (see Figure 6-6).

Figure 6-6

Select your new ringtone from the memory card in your Symbian OS/S60 phone.

Windows Mobile Phone In the Home screen, push START and go to Settings. Select Sounds and you'll see the alert sounds defined for the phone. Click the Ring Tone field to get a list of available ringtones. The list also includes the tunes that you copied to the memory card. Highlight the one you want, and select Done (see Figure 6-7).

Figure 6-7

Select your new ringtone from the memory card in your Windows Mobile phone.

Other Phones Find ringtones, sounds, profiles, or a similar entry from the menu system and select your new ringtone from there.

B) Listen to MP3 Songs and Podcasts

You have the music you want on the memory card. The card is installed into the phone and everything's ready—except for the music player. You have to let the music player know where to find the new MP3 tracks.

Symbian OS/S60 Phone Launch the Music application. You can find it in Media, Fun, My Apps, or in another folder on the phone. Select Update Music Library from the list of options in the main screen. The Music player software will scan the memory card for new MP3 songs and add them to the playlist. When you see the new song titles on the screen, you can start listening to music on your new portable MP3 player (see Figure 6-8).

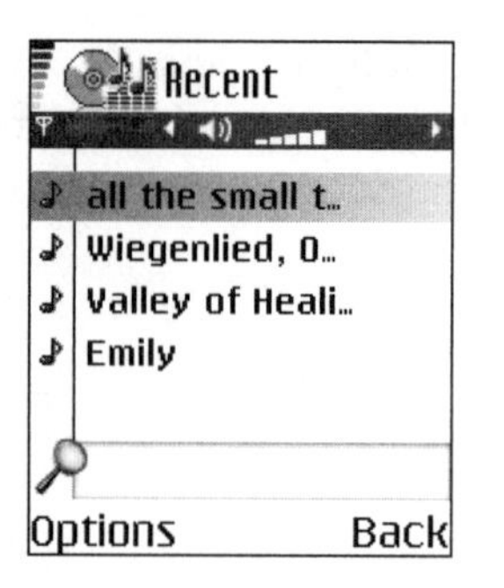

Figure 6-8

Songs stored in the new MP3 player.

Windows Mobile Phone Go to the Home screen and push START. Find the application titled Windows Media and launch it. Open the menu and select Update Library (or Local Content, depending on your phone model). The music player will scan the phone and the memory card for songs it can play (typically, MP3 and WMA tracks). The discovered tracks are listed on the screen. Select a song and start listening to music on your new portable MP3 player (see Figure 6-9).

Figure 6-9

Songs stored on a memory card of your new MP3 player.

Other Phones Find the music player from the phone's menu system. Let the player scan the phone and the memory card for new MP3 songs.

C) Change the Wallpaper

Wallpapers, or Background pictures, are images that you can set to display in the home screen of your device. Earlier, when phone screens were smaller than today, the images were called logos, because phones could only display tiny black-and-white pictures. Now you can define practically any JPG photo, or other common picture format as your background image.

Symbian OS/S60 Phone In the main menu, open Tools | Themes. The check sign in front of the theme title indicates the theme you are currently using. Highlight a theme and then select Options | Edit | Wallpaper | User Defined (see Figure 6-10). The phone will scan the phone memory and the memory card for compatible images. The discovered pictures are listed on the screen. The pictures you copied to the memory card have a small icon after the title. Highlight a wallpaper and choose Select. Go to the Home screen and view your new wallpaper.

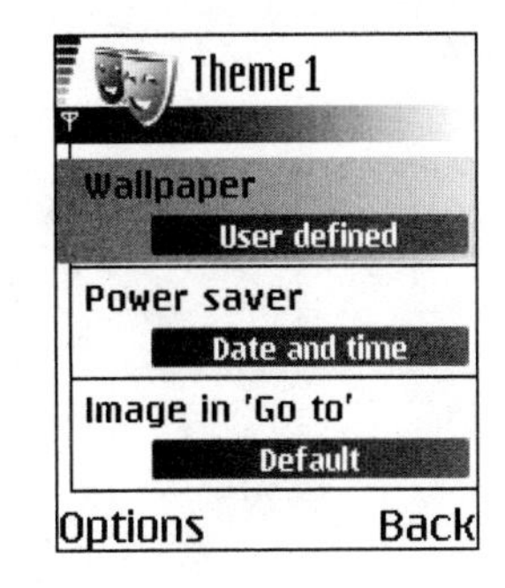

Figure 6-10

Choose a new wallpaper for your Home screen on your Symbian OS/S60 phone.

Windows Mobile Phone Go to the Home screen and push START. Select Settings | Home Screen | Background Image and pick up the picture you want. Click Done when you are ready (see Figure 6-11).

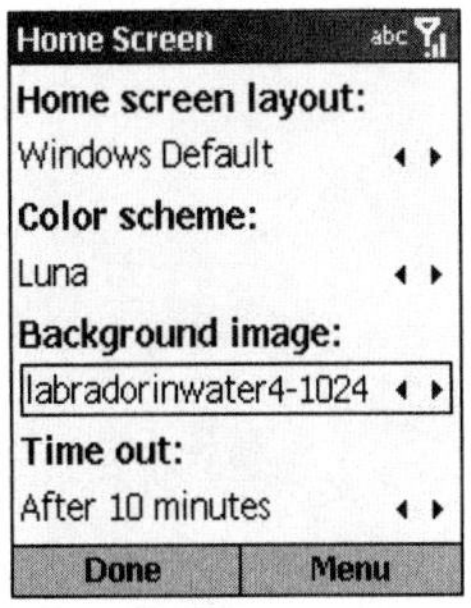

Figure 6-11

Choose a new picture for the Home screen on your Windows Mobile phone.

Other Phones Find an entry titled wallpaper, background picture, logo, or similar from the phone menu to change the wallpaper picture.

D) Watch Videos from the Memory Card

High capacity memory cards can even store (heavily compressed) movies. Whether you want to watch movies, music videos or TV shows on your phone is another matter, but I have enjoyed viewing movie trailers on my phone. Once you have managed to copy the video material on a memory card, you need to tell the media player where to find it.

Symbian OS/S60 Phone Go to the main menu and select Gallery | Video. The Gallery application automatically scans the memory card for new videos. It may list both images and videos, but you can easily recognize videos from the film symbol portrayed over the side of the thumbnail image. Scroll the thumbnails until you find the copied video and select it (see Figure 6-12).

Windows Mobile Phone Go to the Home screen, push START, and find an entry titled Windows Media (or Video Player). Launch it, and open the menu. Select Library | Storage Card. Re-open the menu and select Update Library. Choose My Videos | All Video and pick up the film you want to watch.

Figure 6-12
Music videos, home videos, and perhaps even TV shows and movies on the go.

If your Windows Media software can't play videos (as is the case with older Windows Mobile devices), you must find Video Player application. It will automatically list the video clips it can find in the My Documents/My Videos folder (see Figure 6-13). Highlight the video you want to watch, and select Play.

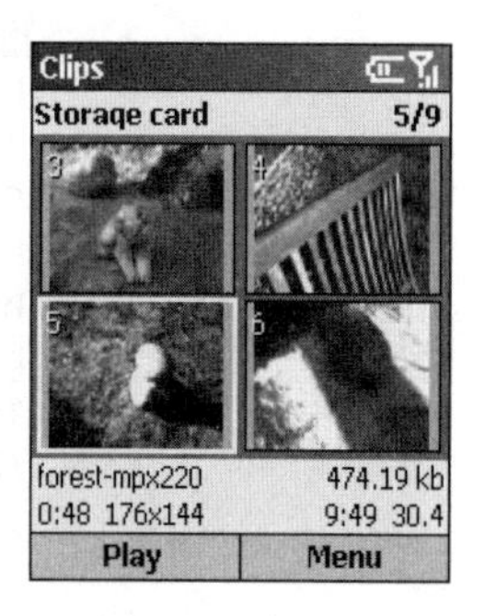

Figure 6-13
You can even watch movies if you have a high-capacity memory card and fully charged battery.

Other Phones Look for a video player in the phone's menu system. Set it to look for videos from the memory card.

E) Install New Software Applications to the Phone from the Memory Card

After you have copied a new mobile application to the memory card, you still have to install the software on your phone. Fortunately, installing software to the phone is easier than a typical software installation on a PC.

Symbian OS/S60 phone Go to the main menu and open the Tools folder. Launch the File Mgr (File Manager) application. Immediately below the File Manager title, you can see small icons. Push the left or right arrow on the keypad and see how the displayed folder and file names change on the screen. You are switching storage spaces between the phone memory and memory card.

When you have the memory card selected, scroll down until you find the folder where you saved the downloaded software product. In my case, I used the folder labeled Software. Open the folder and select the application you want to install (see Figure 6-14). Click on the application, and the phone's installation routine will begin. Follow the instructions until the new product has been installed.

Figure 6-14
Select an application to start the installation routine.

Once installed, you have to find the new application from the phone. It depends on the application and the phone model where the new software is installed. Go to the main menu and scroll all the way down and look for the application. If you can't find it, open Install, My Apps, or Fun folder and scan those folders for the new application.

Windows Mobile Phone Go to the Home screen and press START. Find File Manager and open it. You can now view the top-level folders on the phone. Click the Storage Card. Open the folder where you copied the new applications (in my case, it was Software). Click the application you want to install and the phone's installation procedure will start. Follow the instructions until the procedure is completed (see Figure 6-15).

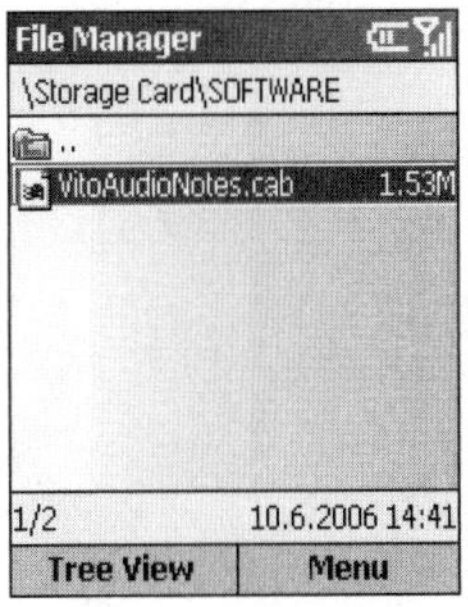

Figure 6-15
Installing Windows Mobile applications from the memory card.

You can run the new application when you go to the Home screen, press START, and browse through the list until you find the new application. On some phone models, Java applications (JAR and JAD files) are placed in the Games & Apps or Midlet Manager folder regardless of the type of applications they happen to be.

Other Phones Search for software installation or a similar entry in the phone's menu system. Follow the installation instructions provided with the phone or the application.

Project 7

Copy Photos, Ringtones, and Other Information from Your PC to Your Phone Using a Wireless Bluetooth Connection

What You'll Need:

- **Bluetooth on your PC. If your PC came without Bluetooth connectivity, it is easy to add it by purchasing a small USB adapter**
- **Bluetooth connectivity on your phone. If your phone didn't come with Bluetooth, you can't usually add it afterwards**
- **Cost: $20–$30 U.S. for a Bluetooth USB adapter for the PC**

Bluetooth wireless technology has been designed to connect devices that are within 30 feet (10 meters) from each other. You can use Bluetooth to transmit information, for example, between a phone and another phone, or a phone and a computer. Bluetooth is also often used for connecting a wireless headset to a phone.

tip *If you are not sure if your PC is equipped with Bluetooth, you can easily verify it. Open the control panel on your PC. Look for an entry labeled Bluetooth (it may be labeled Bluetooth Configuration, or Bluetooth Com, but as long as it says Bluetooth, you're fine). Open the item and check that you can see a list of COM ports assigned for Bluetooth. If you can't, you may have to purchase a low-cost adapter.*

Consider using Bluetooth instead of a memory card for moving information between a phone and a computer when:

- You only want to move a small amount of information between your phone and PC. For example, transferring a few photos via Bluetooth is fast and convenient, but copying MP3 songs is rather slow and requires patience.
- You are using a dedicated application, such as synchronization software that relies on Bluetooth for connectivity.
- There is no memory card on your phone.

You can send any information from a PC to your phone using Bluetooth, but the phone can only make use of certain types of files. Some useful and fun items you might want to transfer from your PC to your phone are:

- *MIDI tunes and WAV ringtones.* Many types of ringtones (monophonic, polyphonic, and real tones) are available for download. MIDI is polyphonic music that most modern phones can play as ringtones. You can identify MIDI tunes by the MID extension in the file name. WAV audio is another universal format for ringtones, but it requires more memory space than MIDI. Advanced music phones can play MP3 tracks as ringtones as well.
- *MP3 tracks.* Any phone that features a music player should play MP3 tracks.
- *JPG (or JPEG) pictures.* You can copy new wallpapers, pictures, and photographs from your PC to your phone.
- *Videos in 3GP format.* 3GP is a standard video format that is recognized by many different camera phone models. Phones that come with Windows Media Player or RealPlayer software are able to play other video formats as well.
- *Software applications.* Using Bluetooth, you can install .SIS or .SISX applications to Symbian OS phones, .CAB applications to Windows Mobile phones, and Java (.JAR and .JAD) applications to most modern phones, including Symbian OS and Windows Mobile devices. In order to install other than CAB applications to a Windows Mobile phone, use the ActiveSync software.

Step 1: Activate Bluetooth on Your Phone

Many phone models display the Bluetooth symbol on the home screen when the connection is active (see Figure 7-1). Even so, you have to verify that the phone can be discovered by other devices via Bluetooth.

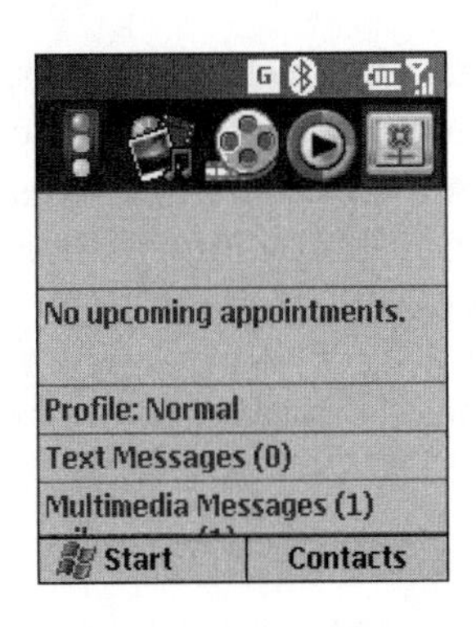

Figure 7-1

This phone is displaying the Bluetooth logo (the symbol that looks like the letter B) at the top of the screen when wireless connectivity has been activated.

Symbian OS/S60 Phone Go to the main menu and open the Connect folder. Find an entry titled Bluetooth and open it. You'll see the current settings for Bluetooth. If the value for Bluetooth is Off, highlight it, and push the selection key to set the value to On. Also, set the phone visibility to Shown to All (see Figure 7-2).

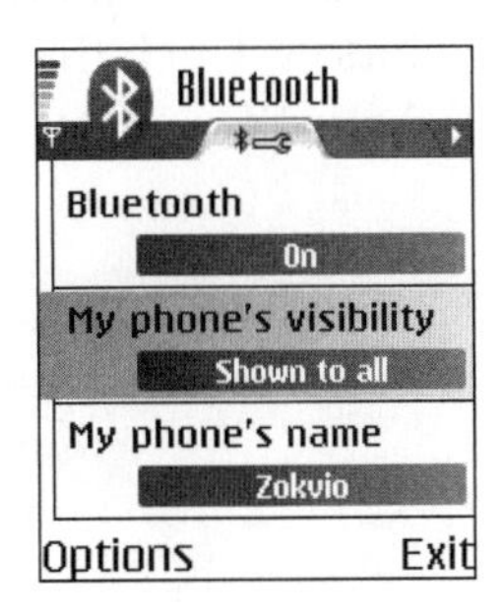

Figure 7-2

Make sure that Bluetooth is set to On and the Phone Visibility is set to Shown to All when you want to receive information on your Symbian OS/S60 phone via Bluetooth.

tip *For security reasons, you can hide your phone's Bluetooth name from other devices. If you do that, no one can find your phone via Bluetooth. There is an exception: connections that you have accepted earlier have already been paired with your device. These devices can connect to your phone, even if the Bluetooth setting doesn't show your Bluetooth name (phone visibility is Hidden).*

Windows Mobile Phone Go to the Home screen and push START. Select Settings. Look for Bluetooth in the list of items (note that the list may be several pages long). If you can find Bluetooth, open it and change the Bluetooth mode to Discoverable (see Figure 7-3). If you don't see Bluetooth in the Settings, open Connections. Then, select Bluetooth and set Bluetooth to Discoverable. Push DONE.

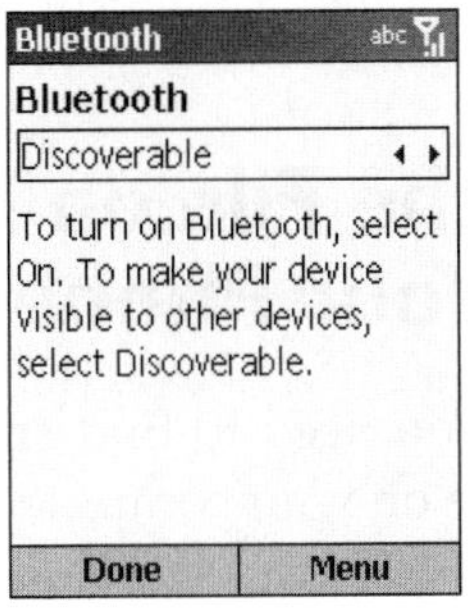

Figure 7-3

Set Bluetooth to Discoverable on your Windows Mobile phone when you want to receive files via Bluetooth.

The value Discoverable means that your phone can be detected when other devices are scanning for Bluetooth connections. The value On means you can transmit information, but other devices won't see the name of your phone. Even in this case, you can still copy information between devices that have earlier been connected (paired) to your phone.

Other Phones Find Bluetooth from the menus and activate it in Discoverable or Visible mode.

Step 2: Activate Bluetooth on Your PC

Now, let's make sure Bluetooth is working on your PC. Go to the Start menu and launch Windows Explorer. Click Folders at the top of the Windows Explorer window. Scroll down the left frame until you find Bluetooth. Click the Bluetooth title.

You should see a new menu item for Bluetooth in the Windows Explorer menu at the top of the window. Click the Bluetooth menu entry and select Device Discovery (see Figure 7-4). Your PC will scan for Bluetooth devices in the range.

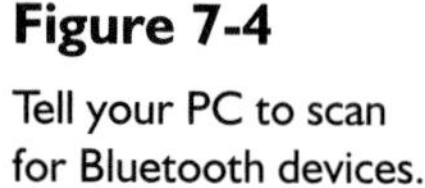

Figure 7-4

Tell your PC to scan for Bluetooth devices.

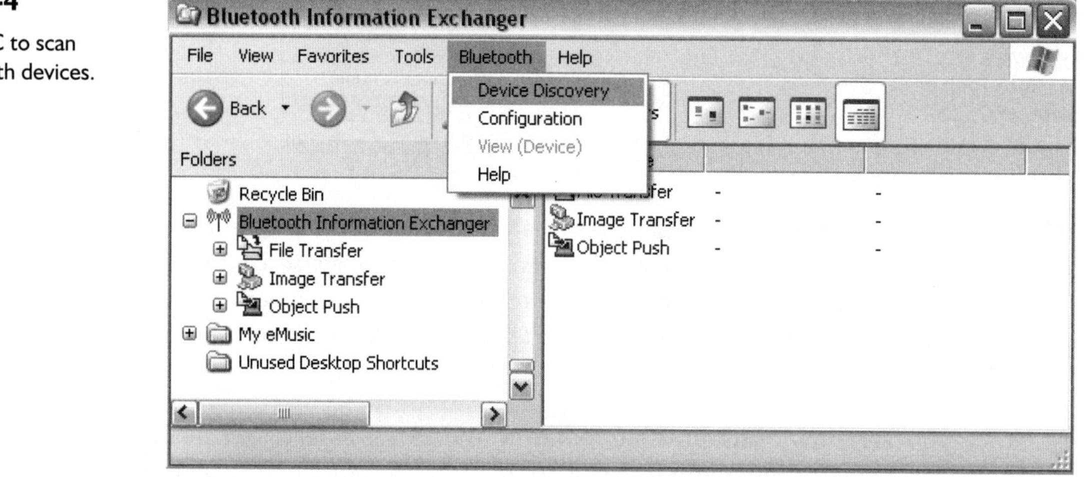

If your PC can't discover any devices, return to step 1 and make sure that Bluetooth on your phone is activated and it is showing its name to other devices. Also, keep in mind that a Bluetooth signal carries about 30 feet (10 meters), but walls and heavy objects decrease the signal coverage.

Step 3: Locate the File on Your PC and Send It to Your Phone

In this step, you need to find an object, for example, an MP3 song, JPG picture, or MIDI ringtone on your computer and send it to the phone. You may be used to playing your MP3 tracks using music player software and viewing your pictures using photo album software on your PC, but usually these applications won't let you send files to other devices. That's why you need to use Windows Explorer for the task.

Open the Start menu on your PC and launch Windows Explorer. Click Folders at the top of the window. If you haven't created your own folder structure on the hard disk, browse folders titled My Pictures, My Music, and My Videos for files you might want to transfer to the phone.

If the Bluetooth software on your PC is rich in features, you can use the following easy technique to send files to your phone. Only some Bluetooth products can do this, but try it out: move your mouse pointer over the file you want to send and right-click. A pop-up menu opens up. If you can find an entry for Bluetooth in the menu, just select your phone from the list and you're done (see Figure 7-5).

Figure 7-5

The easy way to send a file via Bluetooth.

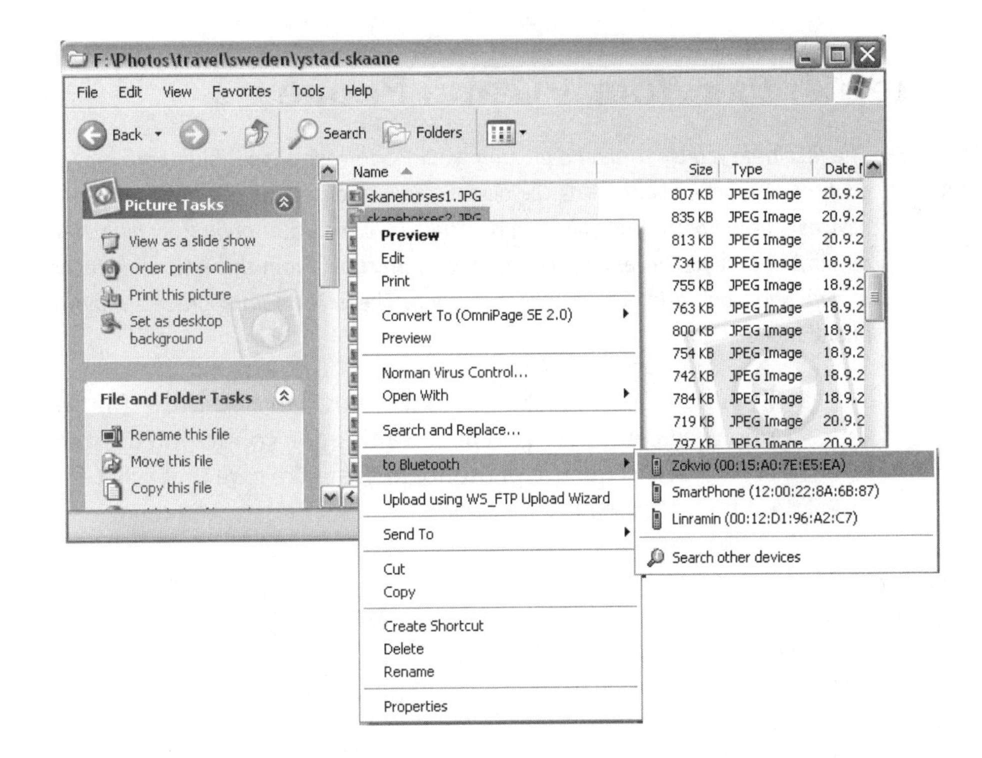

The following technique works with all Bluetooth software, but it requires a few steps to complete. Use Windows Explorer to locate the file you want to send. Once you have discovered the file, list the contents of the folder in the right frame of the Windows Explorer window. Find Bluetooth in the Windows Explorer's left frame, and click the plus sign in front of the title to display its subfolders.

For example, the PC that I'm using displays the discovered Bluetooth devices under the File Transfer folder in the left frame. Click the plus-sign in front of the File Transfer to expand the list. This may vary from one PC to another, depending on the vendor of your Bluetooth software.

You should see the name of your phone in the left frame. Use your mouse to grab the file from the right frame and drag and drop it into your phone in the left frame (see Figure 7-6).

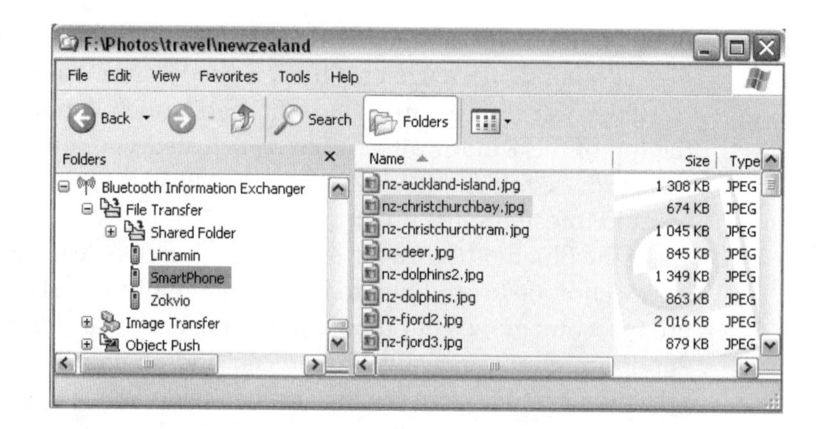

Figure 7-6

Drag and drop a file into the phone to send it via Bluetooth.

Step 4: Pair Your PC and Phone

note *The first time you try to send a file from your PC to your phone, Bluetooth may ask for your confirmation before connecting the devices. You have to think of a code and enter it into both devices. In Bluetooth terminology, this is called pairing. After pairing is completed, the devices remember that you have once allowed the connection. From there on, they regard the connection as safe and won't ask for your reconfirmation. The exact pairing process may vary slightly by phone model.*

Symbian OS/S60 Phone When you have sent the file from your PC, view your phone screen. If the phone is waiting for you to enter a code (Passcode or Passkey), type in any sequence of numbers. Just take care that you can enter the same sequence on your PC (see Figure 7-7).

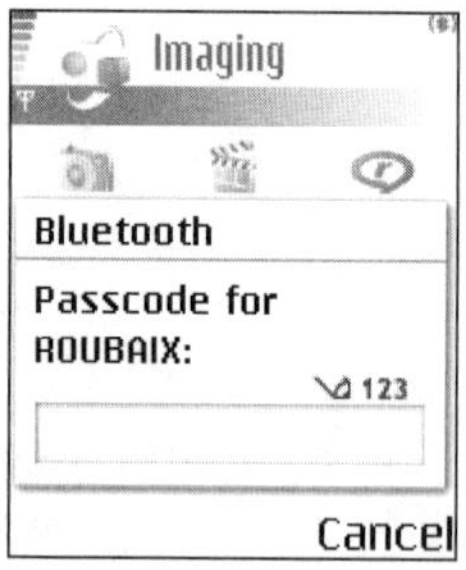

Figure 7-7

Enter a code into your phone for initializing the Bluetooth connection.

After typing in the code on the phone, view your PC screen. The Bluetooth software on the PC is asking for a code. Enter the same sequence of numbers here as you entered into the phone (see Figure 7-8).

After pairing is completed, the phone may ask for your approval for receiving a particular file. Press yes to receive the file.

Windows Mobile Phone Depending on the Bluetooth connection settings, your phone may ask for you to enter a code for confirming the file transfer. Type in any numeric code (Passkey). Just make sure that you can enter the same code on your PC.

Figure 7-8

Enter the same code into your PC.

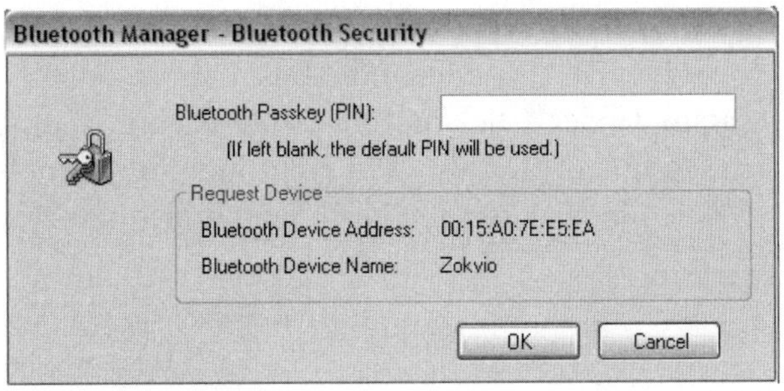

After typing in the code on the phone, view your PC screen. The Bluetooth software on the PC is asking for a code. Enter the same sequence of numbers here as you entered into the phone.

Some Windows Mobile phone models simply ask you to accept or reject the file sent from a PC (see Figure 7-9). If it is the file you want, press Yes.

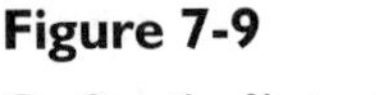

Figure 7-9

Confirm the file transfer.

Other Phones Depending on your phone's Bluetooth connection settings, it may ask for a code (often referred to as Passcode or Passkey). Type in any numeric code. Just make sure that you can enter the same code on your PC. After typing in the code on the phone, view your PC screen. The Bluetooth software on the PC is asking for a code. Enter the same sequence of numbers here as you entered into the phone. Alternatively, some phone models may simply ask you to accept or reject the file sent from a PC.

Step 5: Save the Received File on Your Phone

When you accept to receive a file, it is saved on your phone. Still, there's one more thing for you to do: make sure that the received file goes into a folder where the phone's music player, ringtone selector, or wallpaper chooser can find it.

A) Select Your New Ringtone

Once you have successfully received and saved a new ringtone on your phone, you need to tell the device that you want to use the tune as your new ringer.

Symbian OS/S60 Phone The received ringtone is saved in the phone's Inbox. You can view the contents of the Inbox when you go to the main menu and open the Messaging folder. Click on Inbox (see Figure 7-10). Open the message titled Bluetooth. If the received file is a valid ringtone, the phone's Music Player will play the tune. In Music Player, open the Options menu, and select Set as Ringing Tone.

Figure 7-10

You can find all files received via Bluetooth in the Inbox.

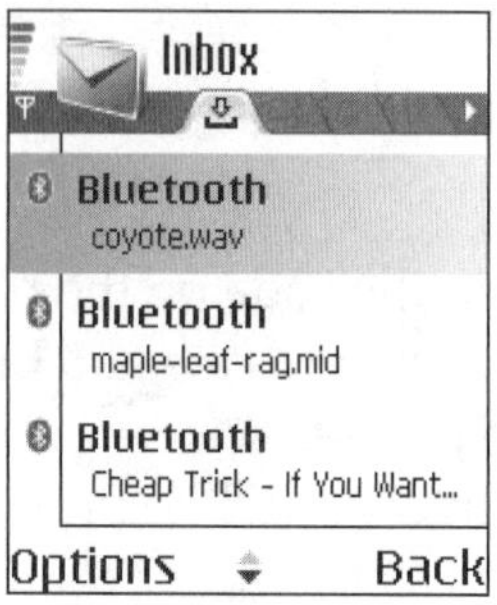

Windows Mobile Phone Depending on your phone model, the received ringtone can be found in a messaging folder, or it has automatically been saved in the phone memory. Launch File Manager. Open the folder \My Documents where you should see the new ringtone. Look also inside the folders FTP or My Music that are subfolders for My Documents.

When you locate the file, click on the received ringtone. An alert screen is displayed, but don't be disturbed. Push MENU and select Save Sound (see Figure 7-11).

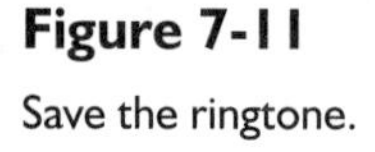

Figure 7-11

Save the ringtone.

There may be differences between phone models where they save different types of files. Navigate folders in the phone memory with File Manager in order to find the tune. When you find it, and want to be sure that the ringtone settings can also find the tune, move the file to \Application Data\Sounds folder.

Next, activate the new ringtone. Go to the Home screen and push START. Select Settings | Sounds. Choose your new tune from the ringtone list. Press Done.

Other Phones Play the received tune. Open the music player menu. Look for a menu item that lets you save the new file as a ringtone.

B) Choose Your New Wallpaper

Many phones can display pictures in the background of their main screen. These pictures are called Wallpapers or Background pictures.

Symbian OS/S60 Phone View your phone's Inbox. Open the picture message received via Bluetooth. Press OPTIONS and select Set as Wallpaper (see Figure 7-12). If you want to store the picture for later viewing, select Save and the picture will be saved in the Gallery.

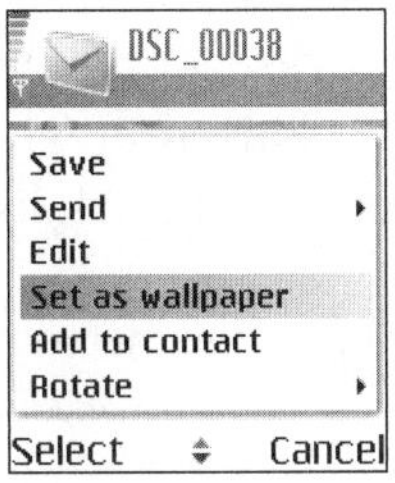

Figure 7-12

Activate your new wallpaper on your Symbian OS/S60 phone. It will be displayed on the Home screen.

Windows Mobile Phone Launch Photo Album (if you don't have Photo Album application on our phone, use File Manager). If you can't see the received picture in the album, you have to change the folder. Open the menu and select Change Folder. Highlight the folder \Storage\My Documents and press ok. When you find the received picture, highlight it and open the menu. Scroll down until you find Apply as Wallpaper. Select it and choose an option to adjust the picture size for the screen (see Figure 7-13).

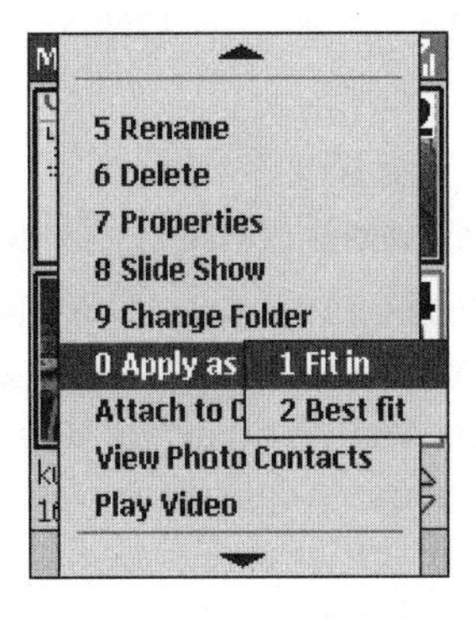

Figure 7-13

Use the Photo Album on your Windows Mobile phone to set the received picture as your new wallpaper.

tip *A technique that works on all Windows Mobile phones requires the use of File Manager. Launch File Manager. Locate the received picture in the \My Documents folder or one of its subfolders. Using File Manager's Move feature, move the image to \Storage Card. Go to Home screen and press Start. Open Settings | Home Screen | Background image. Pick up your new background picture from the list and press Done.*

Other Phones Open the received image. Open the menu and look for a menu entry that lets you set the picture as the new wallpaper.

C) Play Received MP3 Songs and Podcasts

The received MP3 tracks and podcasts are automatically saved in the phone, but the music player may not know it before you update the phone's music library.

Symbian OS/S60 Phone Check the Inbox on your phone. Open the message with an attached MP3 file. The Music Player will start and play the track. Open the Options menu in the Music Player and select Save. If you have a memory card on your phone, save the song on the card.

Next time, when you want to listen to the received song, launch the Music Player application. The application may not realize that new tracks have been imported to the phone. In this case, you have to update the music library by opening Options and selecting Update Collection. The Music Player will search for new tracks and add the discovered songs to the track list.

tip *Once you have saved the MP3 on the memory card, remove the file from the Inbox—it is taking a lot of memory space, and you already have saved the song in another place.*

Windows Mobile Phone Use the Windows Media application for playing audio. Go to the Home screen, push START and find the Windows Media player. When the application starts up, it lists the tracks it has discovered from the phone. Highlight a song and push PLAY.

If you didn't find the new received song in the music library, open the Windows Media application Menu and select Update Library. The player will scan the phone memory and memory card for all types of music files. After the music library has updated itself, select My Music | All Music. Then, pick up the song from the track list.

Other Phones Open the received file and let the phone's music player application play the song.

D) Watch the Received Video Clips

Symbian OS/S60 Phone View the Inbox on your phone. Open the Bluetooth message with a video attached. When you click the message, RealPlayer will launch and play the clip. While you are in RealPlayer, open the Options menu and select Save. If you have a memory card in your phone, save the clip on the memory card to save the phone's internal memory space.

Later, when you want to watch the video, the easiest way is to use the Gallery application. Go to the main menu and launch Gallery. Select Video and find the clip. Push the selection key to watch the video. Remember to remove the video clip from the Inbox—it is taking up plenty of extra space now that you have saved the clip in another folder.

Windows Mobile Phone Go to the Home screen, push START and open the Windows Media application. Open the Menu and select Update Library. After the software has scanned the phone for new videos, select My Videos | All Video. Select a video you want to watch from the list.

If the Windows Media application on your phone doesn't recognize videos at all, launch an application called Video Player. Find your new video from the thumbnail album that the application displays on the screen. If you still can't find the video clip, launch File Manager and view the contents of My Documents folder and its subfolders. Click the file to watch the video.

Other Phones Open the received video clip. View the phone menu, and save the clip in the phone memory or memory card.

E) Install the Received Software Application

Although you have successfully received and saved a new application on your phone, you still have to install it.

caution *Before you install any new software on your phone, keep in mind that some harmful applications for phones have been discovered. These applications can spread via Bluetooth as well. Be careful and install new applications only from sources you trust.*

Symbian OS/S60 Phone List the Inbox on your phone. Open the Bluetooth message with the application attached. The installation routine will start. Follow the instructions to install the application.

Once the installation is completed, you can run the installed product from the main menu or the Fun, My Apps, or Install folder.

Windows Mobile Phone Launch the File Manager application. Open the folder My Documents and look for the received application. Some phone models may save the received file in a subfolder, so take a look into them as well. Once you find the received application, open it, and the installation procedure will start. Follow the instructions.

You can run the installed application by going to the Home screen, pushing Start and locating the new program in the menu.

Other Phones Typically, when a phone receives a Java application (.JAD or .JAR file), it automatically kicks off the software installation routine. Follow the instructions. After the installation is completed, you can find the new application in the phone menu. Some phones store Java applications in folders, like Midlet Manager, Games, or Apps, regardless of the type programs they are.

Project 8

Download New Applications to Your Phone

What You'll Need:

- **A phone that can run Java applications is the minimum requirement. Most modern mid-category and high-end phones are compatible with Java. Symbian OS and Windows Mobile powered devices can run Java applications as well as other software**
- **Internet browser software on the phone. If you download items directly to the phone from the Internet, you must have a browser that can display web (HTML) pages**
- **A data cable and ActiveSync software: Windows Mobile phone owners should have data cable and ActiveSync software installed. They are typically included with the phone**
- **Cost: $0–$25 US per downloaded application**

Many camera phones and music phones are powered by advanced operating systems, such as the Symbian OS, Windows Mobile, Palm OS, or Linux. They introduce features familiar from computers to mobile devices, like e-mail, Internet access, and multimedia. What really sets an advanced phone apart from an ordinary phone is the capability to download and install new software products that enhance the device with new features—ones that weren't there when you opened the package. For example, you could download a training program that keeps track of your fitness and calorie consumption, or software that displays a map for a city you are visiting.

tip *Most phones can run Java applications (identified by the .jar and .jad file extensions). Applications for Symbian OS can be identified by the .sis and .sisx file extensions. Windows Mobile applications are denoted by the .cab file extension. Many Windows Mobile applications are delivered as packages that must be extracted on a PC. These applications are installed to the phone using the ActiveSync software on the PC. If you have downloaded such an application (usually identified by the .exe extension), ensure you have the phone's data cable at hand and ActiveSync installed on your PC.*

First, we are going to browse web sites that provide mobile applications for download. Then, you can decide if you want to download the application directly from the Internet to your phone, or do you first download it on your computer and then copy it to your phone.

Step 1: Log on to an Online Store

Launch the Internet browser on your computer to visit a software store. Try the addresses below. Browse the product listings to get an idea what is available.

Stores with a wide range of mobile software products are, for example, GetJar (www.getjar.com), Handango (www.handango.com), My-Symbian (my-symbian.com), and Smartphone.net (www.smartphone.net). Also, Download.com (www.download.com) has made mobile software available.

You can access some of the mentioned stores with the Internet browser software on your phone as well. For example, My-Symbian and Getjar have made it easy to browse their pages on a mobile device.

Some online stores, like Handango, want to identify your phone model before allowing you to browse their product selection. This way, the store can automatically display only products that are compatible with your phone.

tip *You can usually find the make and the model of your phone under the battery. Switch off your phone, open the battery cover, and remove the battery. You'll find a label that specifies plenty of information about your device, including the model and serial number.*

Step 2: Find the Application That You Want to Download

When you are browsing the product selection in a software store, you will notice that often, there is a trial version and a full version available. Trial applications are fully working products, but with some functional restrictions. A trial software may run only for a few days, or it may not let you save any information. Nonetheless, it is useful to be able to try out a product before buying it.

The full version of a product is often delivered in two phases. First, you download the actual product. Second, once you have paid for the product, you get an e-mail message that includes the code that unlocks the software. Typically, this code is tied to a particular device and you can't move the purchased application to another phone after installation.

There are two methods for downloading applications: saving the file to a computer and copying it to the phone from there, or downloading directly to the phone. Both cases are covered.

Step 3: Install Applications by Downloading Them Directly to Your Phone

note *Before proceeding, keep in mind that when you access the Internet with your phone, you must be subscribing to a data communication plan. Your service provider may call it the wireless Internet, e-mail plan, or something else, but some operators require your confirmation to activate the service. This is because you may have to pay extra for the data service.*

tip *If you are unsure about your phone's capability to transfer information quickly enough, the rule of thumb is as follows. If your phone has GPRS (General Packet Radio Service) capability, the Internet access is slow, but you can still download applications. Phones with EDGE (Enhanced Data Rates for GSM Evolution), EV-DO (Evolution-Data Optimized), or UMTS (Universal Mobile Telecommunications System) connectivity are so fast that it is possible even to watch TV or quickly download items from the Internet.*

The possibility to view maps on a phone is a welcomed capability when you are in a city where you don't know the streets as well as in your home town. Wayfinder Earth (www.wayfinderearth.com) is a downloadable map application that we will use as a sample application.

Using the Internet browser on your computer, log on to www.wayfinderearth.com. Follow the Download links until a web page is displayed that asks for your phone number (see Figure 8-1). Type in your phone number, and click the button to confirm it.

Figure 8-1

Fill in your phone number, and the software vendor will send you a text message for downloading the application.

In this case, the download process is exactly the same for all phones. You will receive a text message that includes a link to the actual download page (see figure 8-2).

Figure 8-2

A text message that includes a link to the download page.

Open the message that you have received to your phone. Click the provided link. If your phone doesn't display the web page, open the menu. Select Use Internet Address, or a similar item from the menu. Depending on your Internet browser, the download of the new application begins automatically or the browser waits for you to confirm it.

If Wayfinder Earth is not compatible with your phone, the download won't start at all. You may try another free map application, such as Google Maps. Launch the Internet browser on your phone and go to: www.google.com/gmm. Follow the instructions to download the application. Even if the web page informs that your phone may not be compatible, start the download. Google Maps is a Java (Midlet) application that runs on a wide range of mobile devices.

When the download has finished, the installation should start (see Figure 8-3). Follow the instructions.

Figure 8-3

After the download, the installation should begin automatically.

Once the installation is completed, you have to find the new application from the phone.

Symbian OS/S60 Phone Look for the new application in the main menu or in the My Apps, Fun, or Install folder.

Windows Mobile Phone Go to the Home screen, press START, and browse the application list. If you have trouble finding it, take a look at the Games & Apps or Midlet Manager folder as well.

Other Phones Search the phone menu for installed applications and Java applications.

Step 4: Install Applications to Your Phone from Your Computer

Even though you will be running the downloaded application on your phone, some stores require you to download the product to the computer first. This means that you have to take a few extra steps for managing the download on your computer and then manage the transfer and installation on the phone.

Symbian OS/S60 Phone Let's use Audio Mixer as the sample application for this installation. Audio Mixer lets you record voices that you can use to create new sounds and ringtones with special effects for your phone.

Launch the Internet browser on your computer. Log on, for instance, to www .my-symbian.com and find the application (see Figure 8-4). You can pick up the trial version of the product by selecting Try It. Start the download, saving the file into a folder where you can find it later.

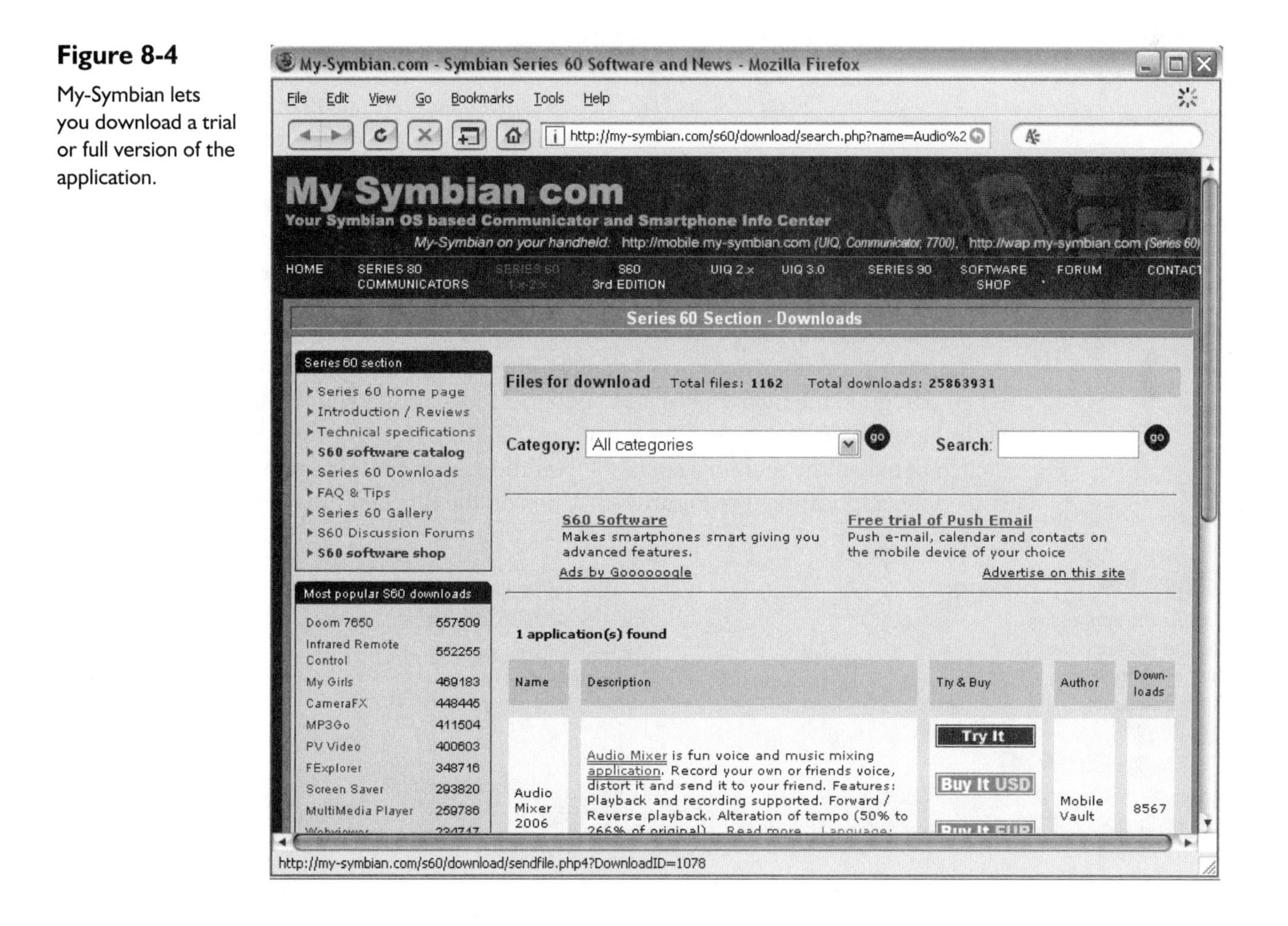

Figure 8-4 My-Symbian lets you download a trial or full version of the application.

Now that you have the application saved on your computer's hard disk, you can use a memory card for installing the application. Insert a memory card into a memory card reader attached to your computer. Copy the application to the memory card using the File Manager software (see Figure 8-5).

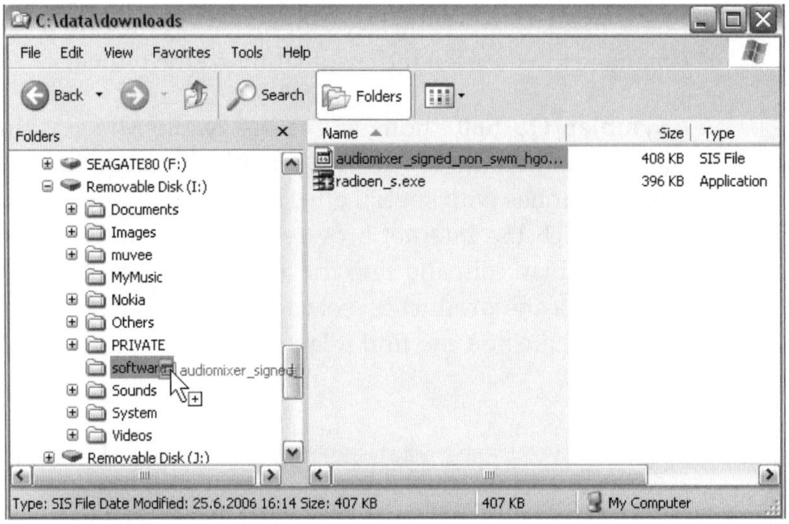

Figure 8-5

Copy the application to the memory card. In this case, I had created the Software folder on the card.

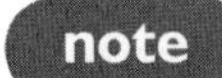

If you have Bluetooth connectivity both on your phone and your computer, you can send the application via the wireless connection to your phone (view Project 7 for details). You may also use the sync software and data cable provided by your phone vendor to install the downloaded software. Connect the data cable from the phone to the PC and start the sync software on the PC.

Remove the memory card from the computer and insert it into your phone.

Launch the File Mgr (File Manager) application on the phone. You can find it when you go to the main menu and open the Tools folder. File Manager lets you view both the phone memory and memory card. Use the left and right navigation keys to switch between the memory spaces. Open the folder on the memory card where you copied the application (see Figure 8-6). Click on the file, and the installation will begin.

Figure 8-6

Locate the application on the memory card and open the file to start installation.

Follow the instructions until the installation is completed. You can find the new application in the phone's main menu or in the My Apps, Fun, or Install folder.

Windows Mobile Phone Listening to Internet radio stations is a lot of fun because of the endless variety they provide. Resco Pocket Radio lets you listen to a few Internet radio stations on your phone. The radio programming is streamed over the mobile network, so you need a generous data communication plan to listen to the shows.

Download the application, for instance, from www.handango.com, and save the file in a folder on your computer where you can find it later (see Figure 8-7).

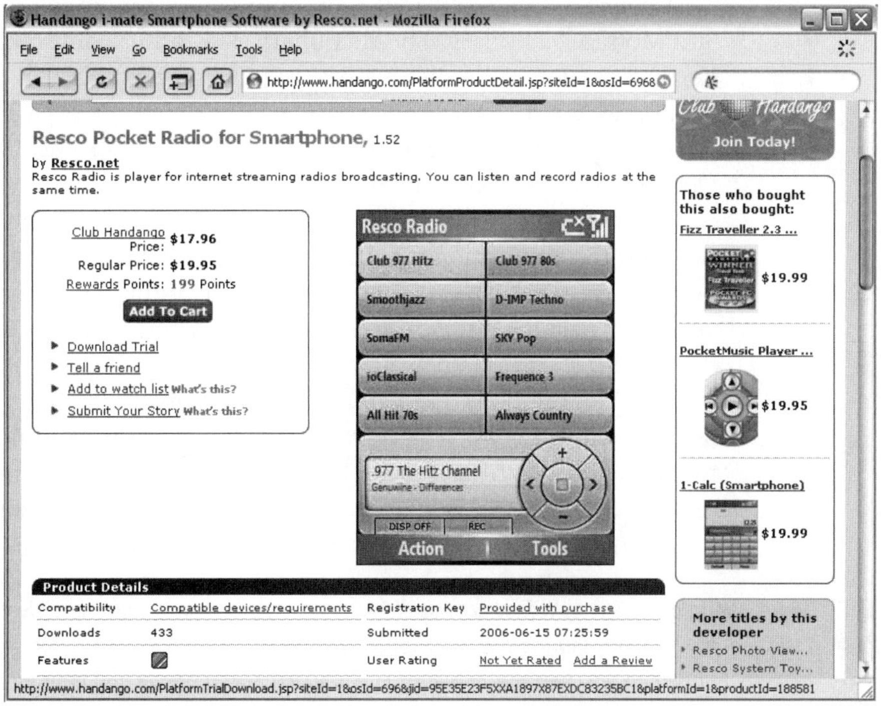

Figure 8-7

Download and save the application on your PC.

Installing a software product delivered as an EXE file requires the use of Microsoft ActiveSync and a data cable. Connect the data cable between your phone and the PC. If ActiveSync doesn't start automatically when you connect the cable (and your phone is switched on), launch ActiveSync on the PC.

Using the File Manager on your PC, find the downloaded file. Double-click the application file for the extraction process to start. ActiveSync will copy the application to your phone, but won't install it. You have to accept the installation by confirming it on the phone (see Figure 8-8).

Figure 8-8

Accept the installation because you know where you got the product and can trust the supplier.

Follow the instructions until the installation is completed. You can find the new application when you go to the Home screen, push START, and browse the list of applications in the menu.

Other Phones A dictionary that's always with you is handy for anyone who occasionally needs to write without the help of a spell-checker. Let's download a Java software for your Java compatible phone.

Using the Internet browser on your computer, go to www.getjar.com. Search for Sun Mobile Dictionary. The dictionary is a rather comprehensive (for a mobile dictionary) catalog of English words, requiring about 450 KB of free memory space on the phone, but it is worth the space.

Once you find the Sun Mobile Dictionary web page, click the SMD.JAR link. Download and save the file on the hard disk.

If you have Bluetooth connectivity both on your phone and your computer, you can send the application via the wireless connection to your phone (view Project 7 for details). You can also use the sync software and data cable provided by your phone vendor to install the downloaded software. Connect the data cable from the phone to the PC and start the sync software on the PC.

Another option is use a memory card for installing the application. Insert a memory card into a memory card reader attached to your computer. Copy the application to the memory card using the File Manager software on the PC. Insert the memory card into the phone. Find a feature that lets you install new applications, or locate Java from the phone menu. Install the Sun Mobile Dictionary.

tip *Some phone models may refuse to install Java applications from anywhere else but from the Internet. In this case, launch the Internet browser on your phone. Enter wap.getjar.com into the address bar. When you find the application you want, download it and it should automatically install.*

Part II

Challenging

Project 9

Load Up Your Phone with Music from Your CDs

What You'll Need:

- **A computer with a CD or DVD disc drive**
- **A music library application on your computer (We'll be using Apple iTunes and Windows Media Player in this project.)**
- **A phone with MP3 player software**
- **A stereo headset is an essential accessory for listening to music**
- **A memory card or a data cable for transferring songs from the PC to the phone**
- **Your music CDs**
- **Cost: $15–$40 U.S. for a memory card**

Extracting music from your own CDs is an easy and economical way to build up a digital music library. Copying music from CDs leaves you with the original material as a backup in case any unfortunate incident wipes out your digital music collection from the computer's hard disk. You can also be rest assured that upgrading to a new computer or to a new MP3 player won't make your digital music library obsolete, because you'll be storing the music in unlocked, universal MP3 format. Songs saved in MP3 format can be played practically on all digital music devices, including music phones.

note *It is easy to complement your music collection with downloads from music download stores, but keep in mind that only a few phones can play copy-protected tracks downloaded from these shops. Motorola has introduced phone models that are compatible with songs downloaded from the iTunes store, and Nokia and Samsung offer phones that can play songs purchased from stores like Musicmatch, Napster, and Rhapsody. The capability to play copy-protected songs is also known as support for digital rights management (DRM) technology.*

We'll go through this project in two phases. First, you will learn to extract songs from your CDs and save them on your computer's hard disk. Second, you will discover the way to copy, or sync the songs from your PC to your phone.

tip *Although this project deals with CDs, it is possible to convert old vinyl records and C-cassette tapes into MP3 as well. Additional software products and audio equipment is required for connecting the old player to the computer and for recording music from vinyl or tape.*

note *Before you start, make sure you have Apple iTunes 6 or 7, or Windows Media Player 10 or 11 installed on your computer. You can download the product of your choice from www.apple.com/itunes/download (Apple iTunes), or from www.microsoft.com/windows/windowsmedia (Windows Media Player). Windows users already have a version of the Media Player on their computers, but before you start building your music library, check that you have the latest version of the product.*

Step 1: Copy Tracks from Your Music CD and Save Them on Your Computer

In this first step, you'll learn to set up your music library and extract tracks from CDs. Tracks are saved as MP3 files on your computer.

1. Launch the Music Library Software on Your Computer

Open the music library application that you have selected as your digital entertainment center.

Although you can change to another application later, it is a good idea to try and stick to the music library product you start with. If the day comes when you want to change to another application, you can do it without losing your MP3 music, but you may have problems with the copy-protected songs. For example, you can't play songs purchased from the Napster store in the iTunes application or tracks downloaded from the iTunes store in the Windows Media Player.

I used Apple iTunes 6 and 7 and Windows Media Player 10 and 11 products running on a Windows PC for this project. If you are using a different version, you may find that one or two features are different, but usually it is easy to find the right setting or menu entry.

2. Set Up Your Digital Music Library

The following settings affect the way a music library application manages your digital song collection:

- *Audio format* Set MP3 as the default format, because it is compatible with practically all digital music devices, including phones.
- *Audio quality* Set the audio quality to 128 Kbps. You may set higher quality if you have sharp ears and plenty of disk space. The higher the value is,

the better the recording quality will be, but the more disk space the songs will require. The limitations in memory space will become evident on your phone. That's why 128 Kbps is a good compression ratio for MP3, because the quality is considered fine for normal music listening. The MP3 file takes about one megabyte of space per one minute of music.

- *Location of the music library on your computer* For example, you could create your music library on a dedicated hard disk F: or in a folder called C:\Music. When you have a dedicated disk or folder for music, it is easy to back up and share the folder on your home network.
- *File name information* Your music library may grow with time, and for future reference, you should save at least the artist and song title in the file name.
- *CD drive* Usually, the music library software automatically finds your CD drive, but not always.

Let's set up iTunes and Windows Media Player for MP3 music.

iTunes

1. Click Edit in the main menu. Select Preferences | Advanced | Importing (see Figure 9-1).
2. Select Good Quality (128 kbps) from the Setting list.
3. Select MP3 Encoder from the Import Using list.
4. Click the General tab and Change button. Choose a new folder if you want to place your music library in another folder or disk drive.

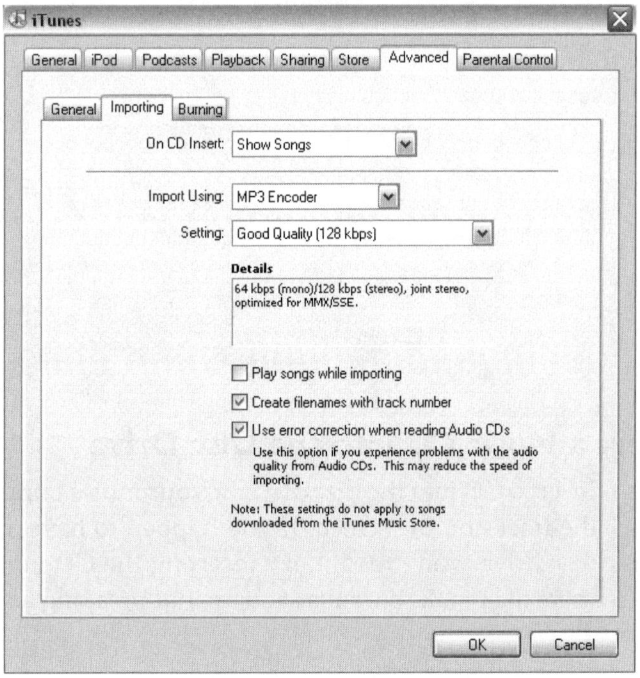

Figure 9-1

Set the MP3 encoder, audio quality, and the music folder location in iTunes Preferences.

Windows Media Player

1. Right-click the small media player symbol at the top-left corner (top-right corner in version 10) of the window. When a menu opens up, select Tools | Options | Rip Music (see Figure 9-2).
2. Rip Music to This Location shows you the folder where the music library currently is located. If you want, you can choose another hard disk or folder. Click the Change button and select a new folder.
3. Click the File Name button and make sure that at least Artist and Song Title are selected. Check more boxes if you want.
4. Rip Settings Format displays the digital format for tracks you import from CDs. Select MP3 from the drop-down list.
5. Move the Audio Quality slider to 128 Kbps.
6. Click the Devices tab. Click Add if the CD or DVD drive on your computer is not listed.

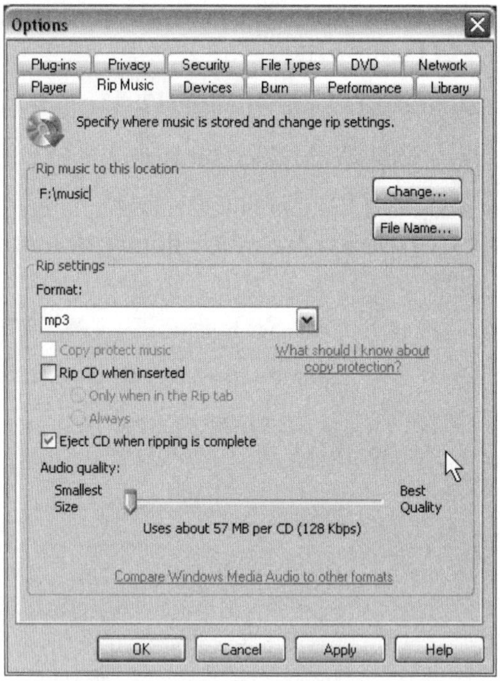

Figure 9-2

Setting up Windows Media Player 11 as the digital music library.

3. Insert a Music CD into the Disc Drive

Insert a CD into the disc drive. If your music library application is running, it will list the tracks on the screen. If you happen to have a copy-protected music CD and the disc drive won't read it, try inserting the CD into a DVD drive and reading it from there (just make sure that you are not breaking any local regulations by doing so).

4. Import Tracks to Your Music Library

When you inserted the CD into the disc drive, the music player software accessed some information from the CD and tried to match it with an online catalog of album titles. The automatic song title discovery requires an active Internet connection. Make sure the titles match the songs on the CD, because from now on, they'll be your only reference to the tracks in your collection. If the track list information is missing, take the time to type in the artist, album, and song names.

iTunes The songs included on the CD are listed in the main window. Check the boxes in front of the titles you want to extract from the CD (see Figure 9-3). Click the Import CD button at the upper right-hand corner (or lower right-hand corner if you are running version 7) of the iTunes window.

Figure 9-3
Import tracks from a CD in iTunes.

Windows Media Player Click Rip on the menu bar at the top. Check the titles you want to copy, and click the Start Rip button (or Rip Music at the top of the window in version 10) that's located at the bottom-right corner of the window (see Figure 9-4).

Once the import has started, wait a few minutes for the process to finish. Then, remove the CD from the disc drive.

5. View Your Music Library

When the import has finished and the new tracks have been saved on the hard disk, you can view and listen to your new digital music collection. You may have to connect loudspeakers or headphones to your computer as well.

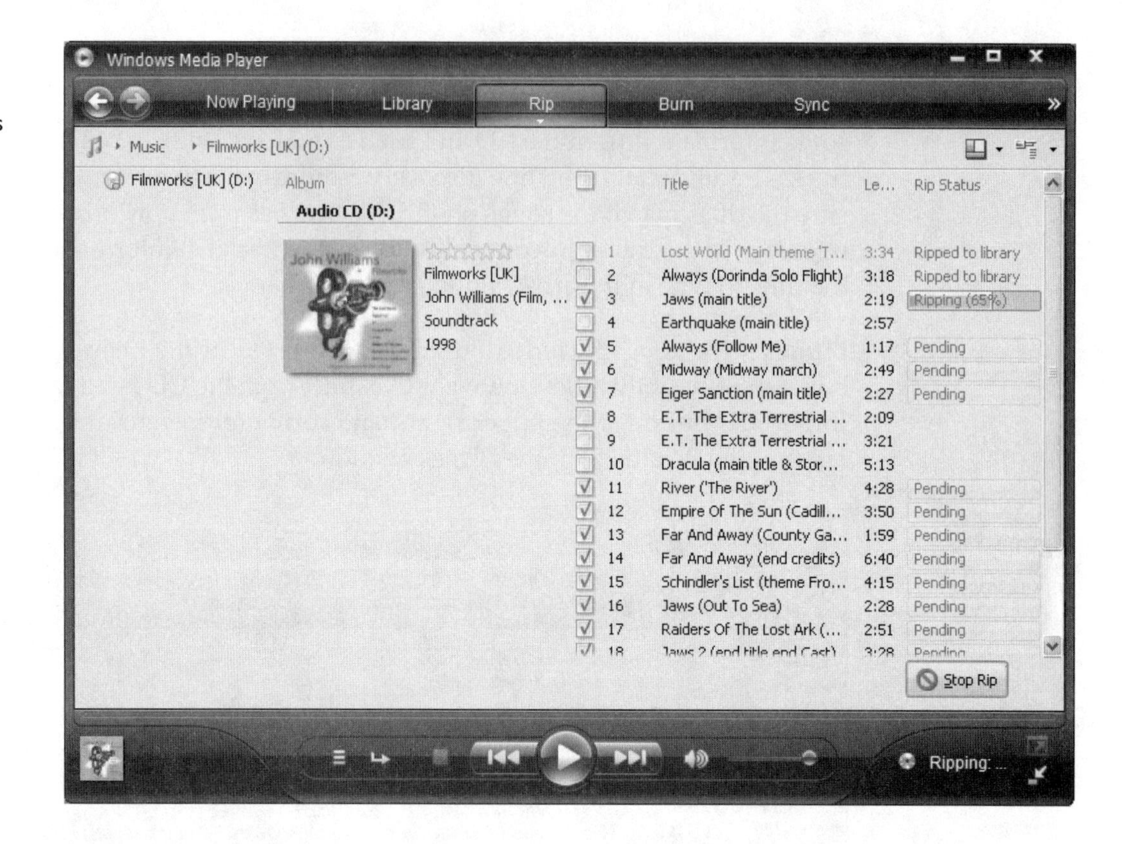

Figure 9-4

Rip songs from a CD in Windows Media Player.

iTunes Click Library | Music in the left pane. Highlight a song and click the Play button at the top of the window.

Windows Media Player Click the Library tab in the menu bar. You can view the tracks in the Library by album, artist, or other criteria by clicking an option on the left pane. Click the large Play button at the bottom of the Media Player window to listen to the song.

For copying additional CDs, repeat items 3 (Insert a Music CD into the Disc Drive) and 4 (Import Tracks to Your Music Library) and keep an eye on your hard disk space to make sure it doesn't run out.

Step 2: Copy Music to Your Phone from Your Music Library

Now that you have established a digital music collection on your computer, it's time to take the next step in the project and copy tracks from the library to the phone. MP3 files are several megabytes in size. Usually, it saves time and effort when you transfer a number of tracks at one go. That's why a memory card and data cable are the fastest and most reliable ways to copy music to the phone. Other possible techniques are wireless Bluetooth communication, infrared, or e-mail.

Phones designed for listening to music often come with extra software that makes it easier to copy tracks to the device. Applications may refer to this feature as downloading, copying, or syncing, but the purpose is the same—to load up a portable device with music. If you received a sync application with your phone, try it out and see if you want to use it for copying tracks.

However, you don't necessarily need any extra software, because you can load up your phone using the equipment you already have. Follow the instructions outlined in Project 6 for using a memory card, or use the technique that we'll go through next—syncing.

Data sync (synchronization) is a process where information is copied between two devices until they have exactly the same pieces of data. In practice, when dealing with music on portable devices, the sync process is usually one way only: from a PC to a portable device. It doesn't usually make sense to sync music from the phone to the PC, because you already should have the songs in your computer.

1. Sync Your Phone with Your Music Library Using iTunes

iTunes software lets you sync music, videos, podcasts, and audio books with iPod audio players and iTunes-compatible music phones. The application can't directly sync with storage media, such as memory cards.

Start iTunes and connect your phone to the computer with the cable that was included with the product. Your phone should appear in the left pane (see Figure 9-5). You can browse the contents of the device.

Figure 9-5

An iPod or an iTunes-compatible phone connected to iTunes.

Choose Edit in the main menu and select Preferences | iPod | Music. You can choose automatic, manual, or playlist sync (see Figure 9-6). In iTunes 7, you can choose between automatic and manual sync when you connect the device to the computer.

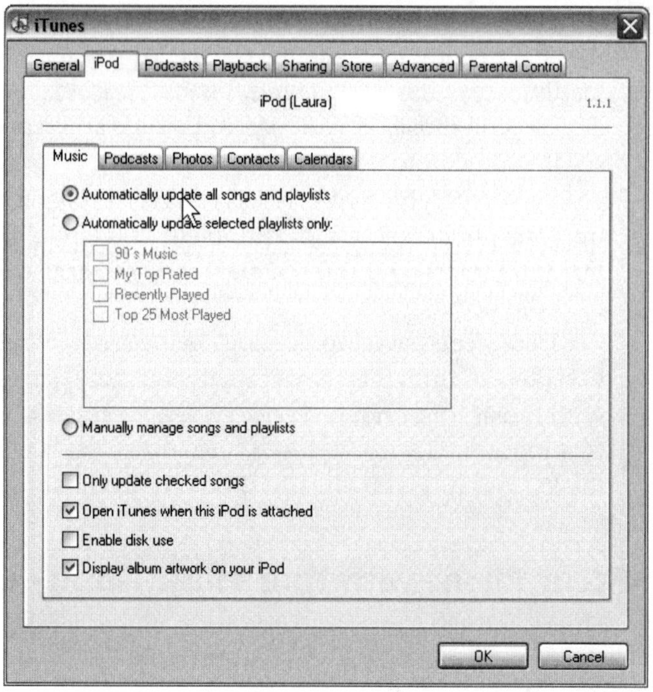

Figure 9-6

Set the sync options for an iPod or for an iTunes-compatible phone.

Since the memory space on a phone is limited, syncing a selected playlist only, or manual sync are the best choices.

If you choose manual sync, it is very easy to copy the songs you want to the phone. While your phone is still connected to iTunes, go to the iTunes main screen. Click Library in the left pane for listing the contents of your music collection. Now, grab any song you like from the list and drag and drop it into your phone in the left pane.

When you want to make space for new tracks on your phone, you can let the sync program manage the memory for you, or you can manually delete songs from the phone. If you manually want to manage the phone memory, list the tracks saved on your phone by clicking the device name in the left pane in iTunes. Right-click a track and select Delete from the pop-up window.

2. Sync Your Phone with Your Music Library Using Windows Media Player

Windows Media Player can sync music, videos, podcasts, and other information with many types of devices, such as MP3 players, portable media players, and even memory cards. When you want to sync your phone using Windows Media Player, you can use a data cable or a memory card as a link between the two. If I can choose which one to use, I use the memory card. That way, I can have memory cards for different purposes. For example, a memory card might include Chinese folk songs, another beats from the Bronx, and one podcasts by travel bloggers.

Connect the Data Cable from Your Phone to Your PC

Launch Windows Media Player. Connect the cable from your phone to your computer.

The first time you connect the data cable from your phone to the computer, Windows may ask your permission to install device-specific software. Accept the installation request. Windows may also ask you to insert a software CD into the disc drive. This refers to a CD that was included with the phone or the data cable (if you purchased the cable separately from the phone). For example, if you have a Windows Mobile smartphone, you would insert the CD with the ActiveSync software into the disc drive.

You can now skip to Create a Sync List, if you are not using a memory card in this project.

Insert a Memory Card into the Card Reader

Launch Windows Media Player. Attach your memory card reader into the computer, or use the built-in memory card slot on a notebook computer.

Remove the memory card from your phone and insert it into the reader. When Windows recognizes the memory card, a pop-up window will appear on your computer screen. Select Manual and click the Finish button to confirm that a memory card has been inserted into the reader.

If Windows Media Player wasn't running when the pop-up window appeared, select the Media Player application from the list.

Create a Sync List

Windows Media Player has a feature called Sync List, which has been designed for portable devices with limited memory space (in practice, for non-hard disk based devices). The Sync List allows Windows Media Player to calculate the required memory space even before copying anything to the device.

Start Windows Media Player, or activate it if it is already running on your computer. Click Sync in the menu bar at the top of the window. In the pane on the right, you should see the label Sync List. In version Windows Media Player 10, click the Edit Playlist link at upper part of the window. Create a new Playlist that will become your Sync List.

List your song collection by clicking Library in the left frame. Find a song that you want to copy to your phone. Grab the title with your mouse, and drag and drop it into the Sync List on the right (see Figure 9-7).

Drop as many tracks as you want into the Sync List, but keep an eye on the memory space. You can follow the amount of remaining memory space on the phone or on the memory card at the top-right pane of the Windows Media Player window.

Save the Sync List you created. Click the Save Playlist button at the bottom right-hand corner (see Figure 9-8).

You can also use the Sync List as a playlist when listening to music on the computer. More importantly, you can use the Sync List as a starting point the next time

Figure 9-7 Drag and drop your favorite songs into the Sync List.

Figure 9-8 Save your new Sync List.

you want to copy songs to your phone. You can find the saved lists under the label Playlists in the upper-left pane of the Windows Media Player window. Highlight the list you want to edit, and click the Edit in List Pane button. The list moves to the right where you can drag and drop items and remove old items.

Start Sync

Click the Start Sync button when you have created a Sync List, and you are ready to copy tracks to your phone. You can follow the progress on the screen (see Figure 9-9).

Figure 9-9 Sync in progress—it may take a few minutes.

note *Windows Media Player will save MP3 files in the top-level folder on the storage media if you used a data cable connection with a Windows Mobile smartphone. If you used a memory card, Windows Media Player will create a folder called Music and subfolders for each artist and album.*

Now, it's time to be entertained by your favorite artists. Start the music player application on your phone, hook up the headset, and enjoy the music on your new MP3 player.

Remove Tracks from the Memory Card to Make Space for New Songs

Eventually, the day will come when you want to update your music collection on your phone. Then, it's time to make space for new tracks. Here's how you do it in Windows Media Player.

When the memory card is in the card reader, or the phone is attached to the PC by a cable, click the Sync button on the Windows Media Player menu bar. Locate your phone or memory card in the left frame. For example, in Figure 9-10, I had named the memory card as MMC_card_64 MB.

Find the albums, artists, or songs on the memory card you want to remove. Right-click the song or album and select Delete (see Figure 9-10).

Figure 9-10

Remove songs from the phone to make space for new ones.

Do not accidentally remove tracks from your computer's music library. Make sure you are viewing the contents of the phone or memory card. The main library located on your computer's hard disk is simply called Library, whereas removable storage media are indicated by a drive letter or by the name of the media.

Project 10

Send Photos, Videos, and Other Information from Your Phone by E-mail

What You'll Need:

- **Dedicated e-mail software on your phone**
- **E-mail account that you can access with a POP3 or IMAP4 communication protocol**
- **MMS application on your phone, if you want to use MMS for sending e-mail. This also requires an activated MMS messaging service from your service provider**
- **Internet (HTML) browser software on the phone, if you access a browser-only e-mail system**
- **Data communication (wireless Internet access) plan from your service provider**
- **Cost: The cost of the data communication plan varies by service provider. Some operators include a number of megabytes per month into their call plans, others charge extra for Internet access. If you use MMS, you may have to pay extra for MMS messages to your service provider**

The possibility to send photos, text, documents, audio, and video clips from your phone by e-mail adds a whole new dimension to your device: the Internet. You can share your photos on moblogs (photo albums designed for camera phones), your home movies on video-sharing sites, and thoughts on personal blogs. Sending a photo greeting to a friend's e-mail box can be a pleasant surprise for your friend as well.

Many phones have not only one, but three options for sending e-mail:

- *A dedicated e-mail application* It is a piece of software running on a phone that allows you to send e-mail messages consisting of pictures, audio, video, and documents.
- *Multimedia messaging (MMS)* Also known as photo messaging or picture messaging, the MMS application lets you send photos, video, and audio directly to other phones and also to e-mail boxes.
- *Internet browser software* You can access your Internet e-mail account with your phone's browser software if your e-mail service features a mobile edition for small screens.

All three options are covered in this project.

Step 1: Set Up and Use E-mail on Your Phone

If you are going to frequently send photos or videos and access e-mail on your phone, configure your phone for full e-mail access (see Figures 10-1 and 10-2).

Figure 10-1

E-mail is one of the messaging applications on a Symbian OS/S60 phone.

Figure 10-2

Messaging on a Windows Mobile phone already includes Outlook e-mail, but you can create more e-mail accounts for other e-mail systems.

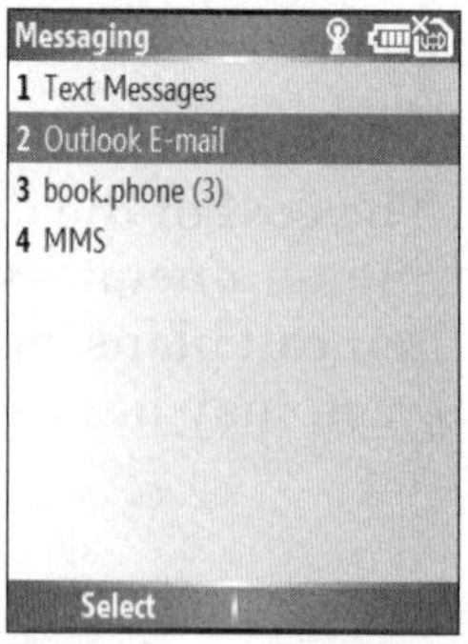

The use of the dedicated e-mail application on your phone allows you to compose new messages, including attachments, without network connection. When you are ready to send the message, you let the e-mail software establish a connection to the network and send the message. For reading new e-mail messages, you let the e-mail software establish the network connection and download only new e-mail titles into your inbox (this can work in many ways, but this is the most common and recommended technique). Then, you mark the messages you want to read, and the e-mail software will download messages along with possible attachments.

For setting up full e-mail access on your phone, you need the following items:

- Dedicated e-mail software on your phone. All Windows Mobile and Symbian phones come with an e-mail application. For other phones, check the phone menu or user guide for e-mail feature.
- An e-mail account that can be accessed from any network and that provides a POP3 or IMAP4 communication protocol. In practice, this means that if you can access the e-mail account using a dedicated e-mail software, like Outlook, Thunderbird, or Eudora on your computer, you can set up your phone to access the same account.

If you fall short with any of the mentioned items, you may use MMS or the Internet browser software for sending e-mail (these are covered later in this project). Phones purchased with a service plan may already be configured for e-mail access. If this is case, you can go directly to Send an E-mail Message.

Let's get started. First, we will set up e-mail access on your phone. Then, we will send a message with a photo attached and see if it goes through to your inbox.

1. Copy the E-mail Account Settings from Your Computer

If you are using Outlook, Thunderbird, Eudora, or another e-mail application on your computer for accessing messages, you can utilize the same account settings for your phone. Open the e-mail application on your computer and find the settings for your account. For example, in Outlook's main menu, select Tools | Accounts and choose your e-mail account from the list. Then select Properties | Servers, and you can view the settings for the servers and e-mail protocol.

Regardless of the e-mail product you are using, look for the following information in the configuration:

- Mailbox type, server type, or e-mail protocol. This should be either POP3 or IMAP4.
- Incoming e-mail server, or IMAP4/POP3 server name.
- Outgoing e-mail server, or SMTP server name.
- User name/login name: this is the user ID you use when you log in to your e-mail account. It may not be the same as your e-mail address.
- Your full e-mail address.

note

Even if you are not using dedicated e-mail software on your computer, it may still be possible to set up e-mail access on your phone. For example, Gmail (www.gmail.com) is typically accessed with Internet browser software, such as Internet Explorer or Firefox. However, Gmail also provides all the required features for accessing it from a phone. If you want to use Gmail, log on to your Gmail account and select Settings | Forwarding | POP to find the configuration information that is needed for this project.

2. Set Up E-mail Access for Your Phone

You have to be careful when configuring your phone. Don't worry; you won't be able to cause any major harm to the device (unless you drop it because your fingers cramp). It's just so easy to misspell those cryptic names—double-check the entries when you are typing them.

note

There is one thing on the phone that may differ from your computer's e-mail settings. That's the outgoing (SMTP) server name. Often, it is a network-dependent name. If your computer has a broadband connection, you use the broadband provider's outgoing (SMTP) server name on your computer. For your phone, you would configure your mobile network operator's outgoing (SMTP) server. Only your service provider can tell you the correct outgoing (SMTP) server name.

tip

If your e-mail system lets you choose between POP3 and IMAP4 e-mail communication protocols, select IMAP4 for your phone. IMAP4 lets you view the titles for all messages, but only downloads those messages to the phone that you explicitly point out that you want to read. Also, the messages are stored on the server until you take explicit action to remove them. This method saves memory space on your phone and makes it fast to access your e-mail account. Some POP3 e-mail systems download all messages to the phone, potentially requiring plenty of memory space. Often, however, it is possible to prevent the download of all messages for POP3 as well—it depends on the features of the e-mail service and on the features of the e-mail software on your phone.

Symbian OS/S60 Phone In the main menu, open Messaging. Press OPTIONS and choose Settings from the list. Select E-mail | Mailboxes and confirm that you want to create a new mailbox. Choose Mailbox Settings and enter the information you collected from your computer's e-mail account into the respective fields on the phone (see Figure 10-3).

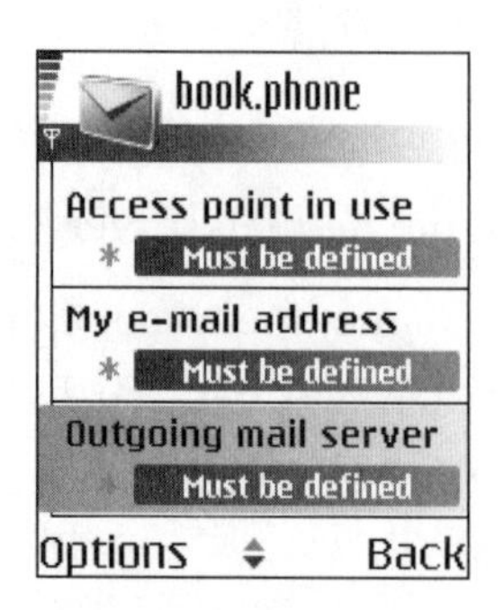

Figure 10-3

Configure the e-mail settings on your Symbian OS/S60 phone.

Windows Mobile Phone In the Home screen, press START and go to Messaging. Choose Outlook E-mail, open the menu, and choose Options | New Account (or Account Setup, Menu, and New). Fill in the e-mail configuration information that you collected from your computer (see Figure 10-4).

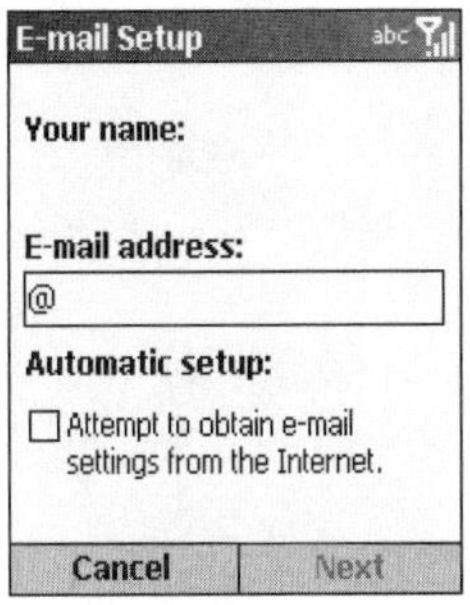

Figure 10-4

Enter the e-mail settings for your Windows Mobile phone.

Other Phones Find e-mail from the phone menu. Enter the information you collected from your computer's e-mail account into the phone's e-mail configuration.

3. Send an E-mail Message

Once you have configured e-mail on your phone, you can try to send and receive a message. Let's take a picture that you have snapped on the camera as a sample item you are going to send, in this case, to yourself.

Symbian OS/S60 Phone Go to the main Menu and open Gallery (or Media, Gallery). Select Images and find the picture you want to send. Select Options | Send | Via E-mail. Type the required information as you would normally do (see Figure 10-5). Send the message to the same address as you have configured on the phone. This way, it is easy to check if the message went through. Once you are ready, select Options | Send.

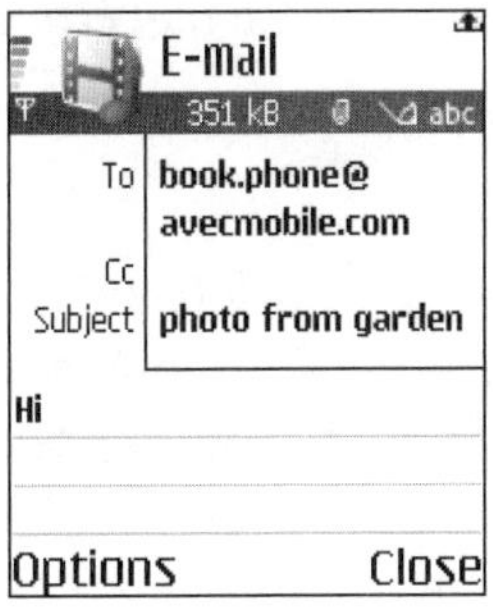

Figure 10-5

Send an e-mail message with a photo attached from your Symbian OS/S60 phone.

Wait a minute to see if the message leaves your phone's outbox. You can find the outbox by opening Messaging from the main menu and scrolling down to Outbox. Open it. If it is empty, messages have been delivered. If the message is still in the outbox, you can try to resend it by selecting Options | Send.

Windows Mobile Phone Go to the Home screen, press START, and open Pictures and Videos (Photo Album). Highlight the picture you want to send. Open the menu; select Send | E-mail. Select your e-mail account from the Messaging list. Type the required information as you would do on your computer (see Figure 10-6). Send the message to the same address you have on the phone. Once you are ready, select Send.

Figure 10-6

Send an e-mail message with an image attached from your Windows Mobile phone.

The message is placed in the outbox to be transmitted the next time you establish a network connection. Well, the next time is now. Go back to the Home screen, press START, and open Messaging. Select your e-mail account. Open the menu and select Send/Receive.

After a minute, you can check if the message has left your phone. Go to Messaging, select the e-mail account, open the menu, and choose Folders | Outbox. If it is empty, all messages have been delivered. If the message is still in the outbox, resend it by opening the menu and selecting Send/Receive.

Other Phones Open an application that can display images (photo or picture album). When you have found the photo you want to send, open the menu, select send by e-mail, and enter the recipient and other information. Send the message and monitor the outbox for its status.

note *If the message is stuck in the outbox, go to Check the Internet Connectivity Settings section later in this project to check your Internet connection.*

4. Check the Inbox

Once the message has left the outbox, you can be assured that the Internet connectivity settings on your phone are fine (you don't have to go through the Check the Internet Connectivity Settings section that follows at all). Assuming that you sent the message to the same e-mail address you have on your phone, there should be a new message in your inbox.

Symbian OS/S60 Phone In the main menu, select Messaging and open your e-mail account. Press OPTIONS and select Connect. Verify your user name, enter your password, and push ok. You'll see the titles of the messages in the inbox (see Figure 10-7). When you click a message title, the whole message is retrieved from the server for you to read.

Figure 10-7

You've got e-mail on your Symbian OS/S60 phone.

If the message contains an attachment (a small paper clip at the top of the screen indicates an attachment), select Options | Attachments. When you choose the attachment, it is downloaded. After the download, select the attachment once more to view it.

Windows Mobile Phone Press START in the Home screen and open Messaging. Select your e-mail account from the list. Push the MENU and choose Send/Receive. Verify your user name and password, and push ok. The titles of new messages will be displayed in the inbox (see Figure 10-8). Select the one you want to read.

Figure 10-8

You've got e-mail on your Windows Mobile smartphone.

If the message includes an attachment, select its link. Push MENU and select Send/Receive. The attachment is downloaded. When the download is over, click the attachment to view it.

Other Phones Open the e-mail application and check the inbox for new messages.

5. Check the Internet Connectivity Settings

If you didn't manage to send or receive e-mail using your phone, check these two things:

1. *Are the e-mail settings correct?* Go back to the section Set Up E-mail Access for Your Phone and verify that all the values are exactly the way you got them from your computer or from your service provider. The usual suspect is the outgoing (SMTP) server name. If you are connected to a mobile network, make sure you have set the outgoing (SMTP) server name of your mobile network operator. If you access the Internet through, for example, your home Wi-Fi, you should set the broadband provider's outgoing (SMTP) server name.
2. *Are the Internet access point settings correct?* Read on to find out.

E-mail settings are separate from the Internet connectivity configuration. This is because Internet access is used by all other communication applications as well, such as the browser and MMS.

You have two options for getting the required Internet access settings into your phone. You may get them from your phone service provider as a configuration text message, or you can manually type the settings in yourself.

I highly recommend you try and get the settings as a text message from your service provider. Use your computer and log on to your phone service provider's Internet page. Look for the information about phone support, settings, phone configuration, or phone setup. If you discover the text message configuration service, log on to the page or enter your phone number and wait for the message. When you receive the message, open it and follow the instructions to save it on your phone.

tip *In addition to network operators, some device manufacturers provide connectivity settings for their phones. For example, Benq-Siemens lets you order settings from their website http://www.benq-siemens.com/.*

If you have a Motorola phone, go to the website http://www.shopmotorola.ca/. Nokia will send you a text message once you have specified your phone model, country, service provider and phone number. Log on to the website http://nokiags.wdsglobal.com/standard?siteLanguageId=118

Internet connectivity configurations for Sony Ericsson phones are available on the website http://www.sonyericsson.com/.

If your service provider doesn't send settings as text messages, you can enter the values yourself. Still, you have to find the correct settings for the Internet connectivity from your service provider's Internet pages or by calling the customer service.

Symbian OS/S60 Phone In order to manually enter the settings, go to the main menu, choose Tools | Settings | Connection and Access Points. Open the Options menu and choose New Access Point. Then, type in the information that your service provider gave you (see Figure 10-9).

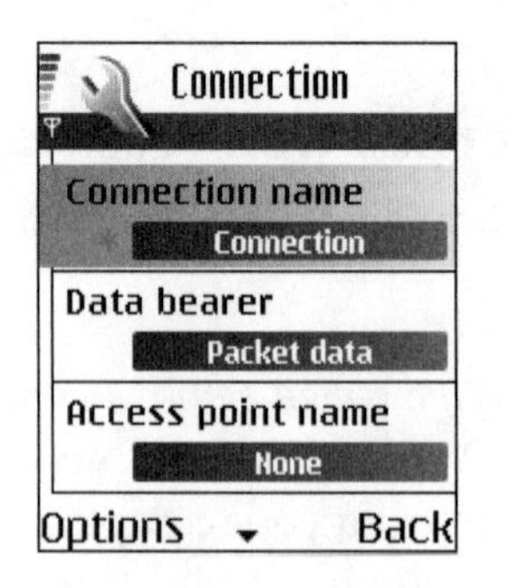

Figure 10-9
Enter the information for the Internet connection on your Symbian OS/S60 phone.

Windows Mobile Phone In the Home screen, press START, select Settings | Data Connections (or Connections and GPRS). Type in the information your service provider specified for you (see Figure 10-10).

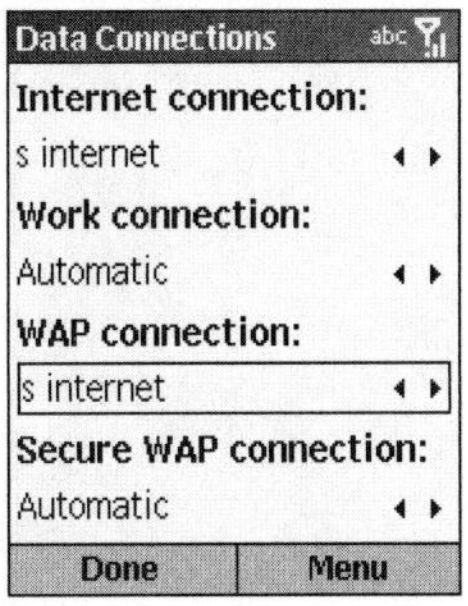

Figure 10-10

Enter the details for Internet connectivity on your Windows Mobile phone.

Data Connections
Internet connection:
s internet
Work connection:
Automatic
WAP connection:
s internet
Secure WAP connection:
Automatic
Done Menu

Other Phones Find Internet connection settings from the menu and enter the information provided by your network operator.

Now, you can retry sending the message from your phone.

Step 2: Use Multimedia Messaging for E-mail

Multimedia messaging service (MMS) allows you to send images, audio, and video clips from your phone to other MMS-capable phones. In addition, it is possible to send MMS messages to e-mail addresses.

To send MMS messages, you need

- MMS feature on your phone
- An activated MMS messaging service from your service provider

note *Before blasting a bunch of multimedia messages out into the world, make sure that you are familiar with your service provider's MMS pricing policy. Some operators charge for each message while others include a number of messages in a monthly plan.*

tip *Make sure the size of the MMS message (for example, an image) is not larger than 50 KB. Many MMS systems limit the size of messages they let through. The maximum allowed size varies by service provider, but it is difficult to know the allowed size before you actually send a message and monitor if it goes through.*

Let's try and send an MMS message from your phone to an e-mail address.

Symbian OS/S60 Phone Open the Gallery application and find a photo that you want to send. View the photo, open the Options menu, and select Send | Via Multimedia. Many MMS services have restrictions on the image size, but the phone software will automatically scale down a large picture.

Here's the thing that turns MMS into e-mail: type in an e-mail address into the To-field (see Figure 10-11). There's nothing else to it. If you want to send the message to a friend who has an MMS-capable phone, type in the phone number or pick it up from Contacts by selecting the To-field. You can also write some text to go with the picture. When you are done, select Send from the Options menu.

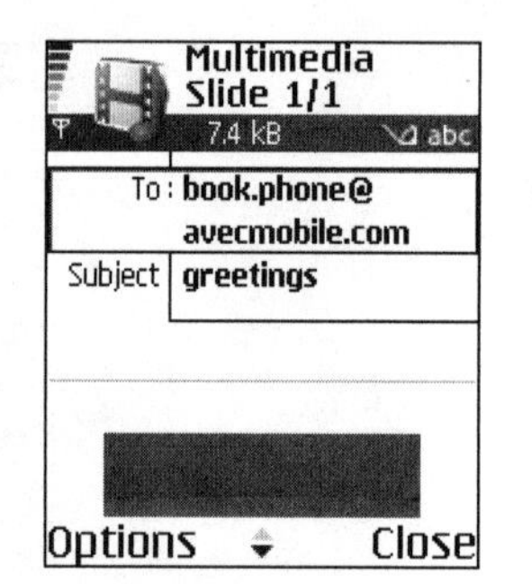

Figure 10-11
Send an MMS message to an e-mail address using your Symbian OS/S60 phone.

Windows Mobile Phone Open the Pictures & Videos (Photo Album) application and find a photo that you want to send. You can check the image size at the bottom of the Photo Album screen when you are browsing photos (or open the photo and select Properties from the menu). Highlight the photo, open the menu, and select Send Via | MMS.

Type an e-mail address into the To-field (see Figure 10-12). If you have saved e-mail addresses in your Contacts list, you can pick up the e-mail address from there. Open the menu, select Recipients | Add, and pick up an e-mail address from your contacts.

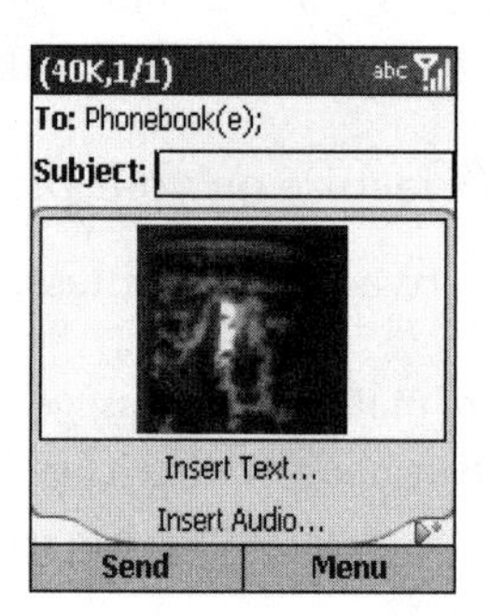

Figure 10-12
Send an MMS to an e-mail address using your Windows Mobile phone.

Other Phones Open the MMS feature on your phone. Remembering potential message size restrictions, attach a photo into the message. Type the recipient's e-mail address into the field where you would enter the recipient's phone number. Send the message.

If you have problems typing the e-mail address, you may have to switch to alpha mode on the keypad. On a Windows Mobile phone, keep STAR (*) pushed down until you see "Abc" at the top bar (or a menu pops up where you can choose letters). On a Symbian S60 phone, press the button with the pen symbol and select Alpha Mode.

Monitor your phone's outbox. If the message is stuck in there, it is likely that you are lacking the MMS settings. You have to get the correct MMS settings from your phone service provider. Look them up on the service provider's Internet pages, or call the customer service and ask them to provide the settings.

note *When you send an MMS to an e-mail address, the recipient will usually see your phone number as the sender of the message instead of your name or e-mail address. It is a good policy to sign your MMS messages so that the recipient can easily recognize the sender.*

Step 3: Use Your Internet Browser Software for E-mail

It is possible to use your phone's Internet browser software for accessing e-mail as well. If the Internet e-mail service you are using only allows access to the account using a browser software, or the other techniques for sending e-mail have failed, you can try the browser. Popular e-mail services, like Yahoo Mail, Hotmail, and Gmail, can be accessed on a phone browser that can display web pages.

note *There is one shortcoming when using a browser for e-mail. Usually, it is not possible to add attachments to outgoing messages, but you can usually read received attachments.*

For accessing your e-mail account with the Internet browser on the phone, you will need:

- *Internet (HTML) browser software.* For example, all Windows Mobile phones come with Internet Explorer which lets you access Web pages. If you don't have Internet browser software on your phone, check if Opera Mini (mini.opera.com) is compatible with your phone. It is a free download.
- *The address of the mobile edition of your Internet e-mail service.* For example, instead of www.gmail.com, you should enter the address **m.gmail.com** for accessing Gmail from your phone. For Hotmail, MSN, or Windows Live Mail, type **mobile.live.com**. If you want Yahoo Mail, enter **mobile.yahoo.com/mail** in the browser of your phone.
- *Correctly configured Internet connectivity settings on your phone.*

Symbian OS/S60 Phone Go to the main menu and open the Web (or Services if you can't find Web) application. Type the Internet address for the mobile version of the e-mail system you are using. Log on to the e-mail account with your user name and password (see Figure 10-13).

Figure 10-13
Gmail on your Symbian OS/S60 phone.

Windows Mobile Phone In the Home screen, press START and select Internet Explorer. Open the menu and select Address Bar. Type in the mobile edition address of the e-mail system you are using (Figure 10-14). Log on with your user name and password.

Figure 10-14

Hotmail on your Windows Mobile phone.

Other Phones Launch the Internet browser software on your phone. Enter the address of the e-mail service you want to access into the browser's address bar.

Learn about Mobile Messaging Costs

Costs are always a factor when using a mobile phone, be it for talking or messaging. The total cost always depends on your service plan. Find out if you have any data communication time or megabytes for data transfer included in your plan, or if MMS messages are bundled with your plan.

E-mail is usually the cheapest option for frequently sending and receiving photos (or any attachments) from a phone. The only cost is for data communication, and the cost accumulates only if you go over your plan's monthly limit. Camera phone images don't easily generate megabytes of data traffic per month (unless you have a megapixel camera and you send everything to your Buzznet, Flickr, YouTube, or another online album). E-mail is a good choice for sending photos to friends and for posting photos and videos on online albums.

MMS pricing is a different story: most service providers charge you for each message you send. They tend to charge a fixed sum for a message, no matter the size. So, why not send a bunch of megapixel photos in one MMS message? Service providers often restrict the size of the message to around 50–100 KB. In practice, that's the size of one or three VGA (640×480) resolution photos. Get to know your service plan's MMS pricing, and you'll be fine with this option as well.

Browsing Internet pages on a phone eats data transfer megabytes that may be included in your data plan. It doesn't make a difference if you use the browser software for e-mail access or for something else. The same pricing policies usually apply both for browsing and for e-mail.

Project 11

Load Up Your Phone with Polyphonic MIDI Ringtones

What You'll Need:

- **You must have phone that can play MIDI audio files as its ringtones. Practically all phones that can play polyphonic audio, can play MIDI tunes**
- **Internet (HTML) browser on the phone for downloading ringtones directly to the phone**
- **MIDI ringtones. You can purchase, find free downloads, or create MIDI ringtones yourself**
- **Cost: $2 U.S. or less per ringtone**

Even if you don't have a fancy music phone that can play MP3 songs, it is usually possible to add melodic ringtones to your phone. Most modern phones can play polyphonic MIDI tunes that can be downloaded from the Internet, or you can create your own ringtones using a MIDI composer application on a computer. This makes it easier for you to frequently change ringtones without paying too much money for new tunes.

Musicians have used MIDI (Musical Instrument Digital Interface) for many years for exchanging notes in digital format between instruments. When polyphonic ringtones were introduced to phones, MIDI quickly became a popular format. You'll have to check your phone's user guide for its MIDI capability, but practically all recent phones can play MIDI tunes.

Since there are plenty of MIDI software products available, and since amateur musicians have made MIDI tunes available on the Internet, ringtone lovers have a wide selection of tunes to choose from. You can download MIDI tunes from the Internet or create new MIDI melodies on computer software. Because MIDI tunes are synthesized melodies, the singer's voice is replaced by an instrument.

MIDI files are typically small. In less than 20KB, you can have the melody and arrangement for a full song. MIDI files download quickly, and they don't take much memory space on the phone. This is in contrast with MP3 and other digital music formats that may take up to one megabyte of storage space per minute of music.

Your phone service provider probably offers a selection of ringtones on its Web pages. However, if you want to discover more variety in offering and pricing, check out ringtone services that are available elsewhere on the Internet. Many of them provide tunes for download independent of your country or phone operator. Some MIDI ringtone download services are, for example, Ringtones (www.ringtones.lt), Midiringtones (www.midiringtones.com), Jamster (www.jamster.com), Partners In Rhyme (www.partnersinrhyme.com/midi), and Polyphonicringtonez (www.polyphonicringtonez.com).

Step 1: Download MIDI Ringtones Using the Browser Software on Your Phone

The easiest technique to get new MIDI ringtones is to download them directly to the phone from a Web page. You simply use your phone's browser software to log on to a service that provides ringtones and download new tunes into the phone.

For achieving this, you need a phone that has an Internet (HTML) or WAP (Wireless Application Protocol) browser application. Try to download a tune by visiting, for example, the Partners In Rhyme Web site (www.partnersinrhyme.com).

tip *If you are unsure which browser you have, here is a rule of thumb: smartphones (typically, devices powered by Symbian OS/S60, UIQ, Windows Mobile, Palm OS, or Linux) come with HTML browser software that can display practically all Internet pages. Other phones, assuming that they have the capability to connect to the Internet, often come with WAP browser software (and they may feature an Internet browser as well). WAP is intended for viewing compact pages specifically designed for phones.*

Symbian OS/S60 Phone Go to the main menu and open Services (Web). Type the address (see Figure 11-1): **www.partnersinrhyme.com/midi** into the address field.

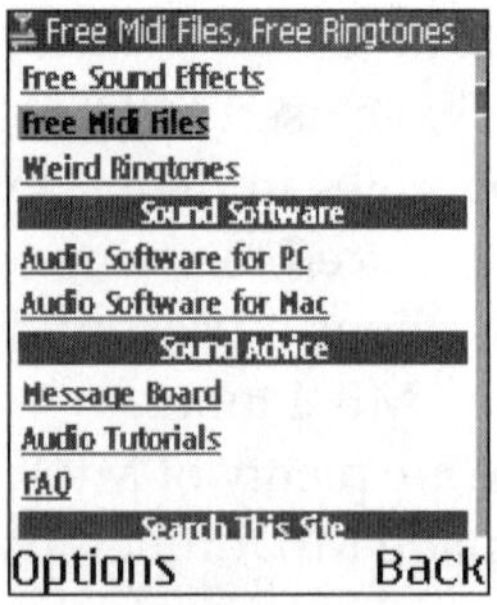

Figure 11-1

Log on to a Web page for MIDI ringtones.

Navigate the pages for the tune you desire. When you select the link of the MIDI song, the ringtone is downloaded. The music player on the phone will automatically launch and play the tune. Open Options and select Set as Ringing Tone (see Figure 11-2).

Figure 11-2
If you are happy with the tune, set it as your new ringtone.

Windows Mobile Phone Go to the Home screen, press START, and launch Internet Explorer. Open the menu and select Address Bar. Type the following address into the address field (see Figure 11-3): **www.partnersinrhyme.com/midi**.

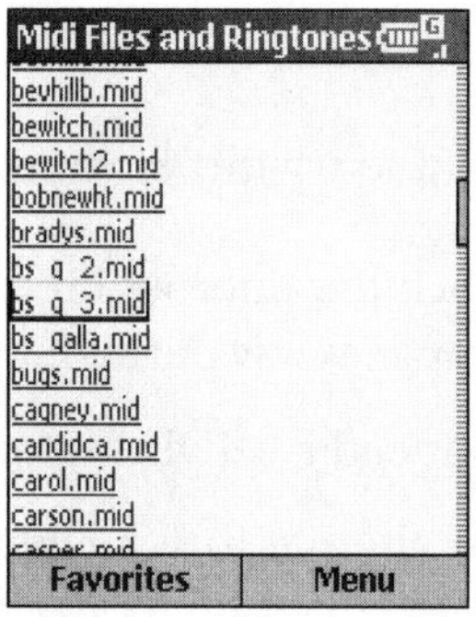

Figure 11-3
Visit a MIDI download service using your phone browser.

Find a tune by navigating the links on the pages. When you click a MIDI file, the phone will ask you to confirm the download. Choose Yes and the ringtone will download. When the download has finished, open the menu and select Save Sound (see Figure 11-4). Some phone models automatically launch the music player application and save the tune.

Figure 11-4
Save the new MIDI ringtone.

The ringtone has been saved, but you still have to set it as the default ringtone. In the Home screen, push START and select Settings | Sounds | Ring Tone. Then, you can view the ringtones stored on the phone and pick up the new MIDI tune.

Other Phones If you have a phone with an WAP browser software, type **http://www.polyphonicringtonez.com/wap** into the address bar. Find your new ringtone and download it (see Figure 11-5). Don't forget to set the tune as your new ringtone in the phone settings.

Figure 11-5
Access a MIDI download site with a WAP browser.

Step 2: Download Ringtones to Your Computer and Copy Them to Your Phone

You can also access MIDI ringtone Web sites using the Internet browser on your computer.

If you want to try this technique for free ringtone downloads, do as follows:

1. Launch the Internet browser on your computer, and Go to a Web page, for example, www.partnersinrhyme.com/midi.
2. Download a MIDI ringtone and save it on your computer's hard disk.
3. Copy the file to your phone from the hard disk. Use a memory card (see Project 2), or Bluetooth connection (see Project 3) for transferring new tunes to the phone from your computer. If your phone came with sync software, you may be able to use that application for copying the new ringtone to your phone as well.
4. Save the new ringtone in your phone.
5. Activate the new tune by going into the phone's ringtone settings and selecting the new tune as your default ringer.

If you purchase ringtones, make sure that you order one ringtone at a time only. Some online stores push monthly subscriptions that let you have more ringtones, but you have to pay for additional ringtones—whether you want them or not.

The most common way to download and pay for ringtones is to send a text message to a special phone number. The service will respond with a link to the ringtone download. The cost will be added to your phone bill, or you pay with a credit card.

Step 3: Create Your Own MIDI Ringtones

You probably have heard your share of lousy ringtones. If you have ever thought that it would be easy to compose better tunes (regardless if you are musically talented or not), give it a try. Install a piece of software on your computer and create a melody. I can tell you it is not easy to compose tunes, but it can be fun and you are guaranteed to get a unique ringtone.

Musicians have used sophisticated software for many years to help them notate and arrange music. Many of these software products use MIDI as the language to tag the music. Fortunately, there's also a wide selection of MIDI software products available that an average computer user can use for composing ringtone masterpieces of their own.

Anvil Studio and MIDI Piano are examples of products that you can download for free and use for composing your own melodies.

Anvil Studio

Anvil Studio is an advanced software product for working with MIDI tunes (see Figure 11-6). It is available for download at www.anvilstudio.com. You can hook up musical instruments to your computer and let Anvil Studio record what is being played. You may also use the piano keyboard on the screen to compose new melodies.

Figure 11-6

Anvil Studio is an advanced tool for working with instruments and arrangements.

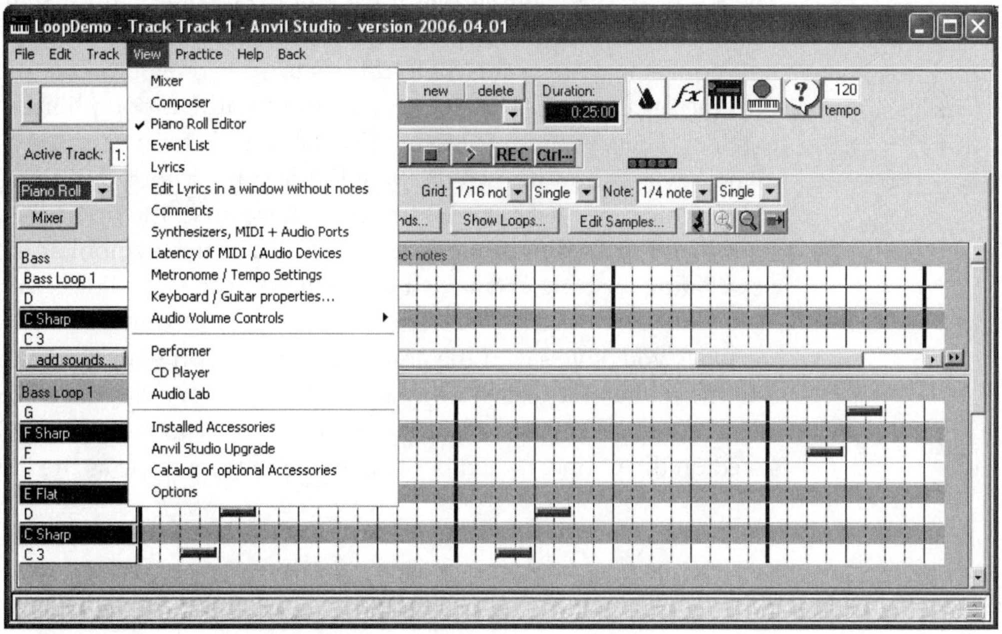

The easy way to start composing is to use an existing melody as a template. Open a tune in Anvil Studio, play it, and find the section you like the best. The intro segment in a pop song often works well as a ringtone. If you like classical music, play the piece until you find the segment with the theme (or any other part you like) and cut that into a 30-second piece to make it a nice ringtone.

When you are ready, open the File menu and select Save Song As. Give your MIDI tune a title and save it.

MidiPiano

MidiPiano is a small, simple application for anyone who knows something about playing piano (see Figure 11-7). It is available for download at www.midipiano.net.

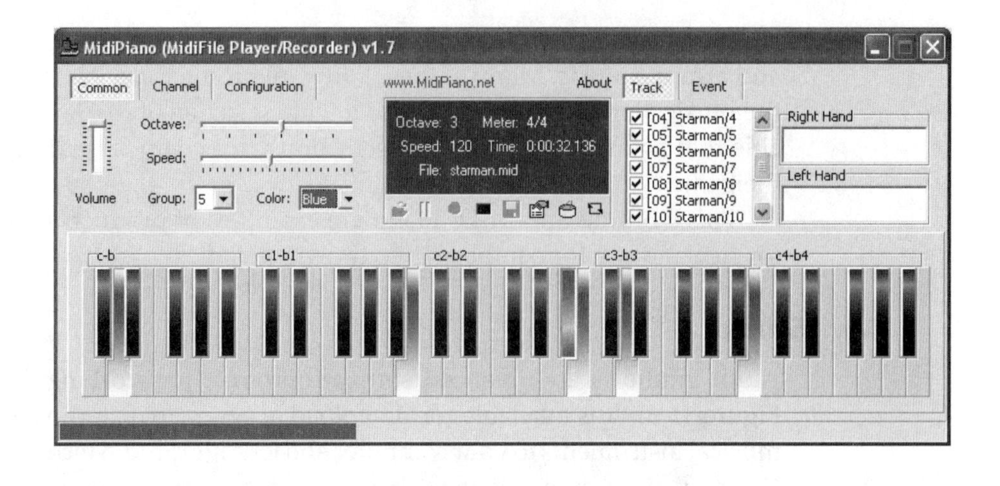

Figure 11-7

MidiPiano lets you create tunes by clicking the piano keyboard on the PC screen.

The software is so simple that it doesn't even come with installation software. You download the application, unzip it, and launch the EXE file to run it.

A quick way to create MIDI tunes is to open one of the sample melodies and play it in the application. This way, you can see how a polyphonic MIDI tune was put together. Then, you can try to modify the sample tune or create a new one. Using the piano keyboard on the screen, play the first track for your new song. If you want a polyphonic tune, just record additional tracks over the first one.

When the new ringtone is ready, click the disk symbol at the center of the window. Name the tune and save it on the hard disk.

Once you have saved the new musical piece on your computer, you have to copy it to the phone. The easiest way is to use a memory card (see Project 2) or Bluetooth (see Project 3) for transferring the tune to your mobile device. Once the ringtone is saved on the phone, remember to activate the new tune as the default ringtone.

Project 12

Share Your Pictures on an Internet Photo Album

What You'll Need:

- **Account on a photo sharing service on the Internet**
- **E-mail access or MMS feature on your phone**
- **Data communication (wireless Internet access) plan from your service provider**
- **Cost: The cost of sending photos to the Internet from your phone depends on your data subscription plan**

You've taken some excellent photos on your camera phone that you are dying to show to the world, or at least, to your friends and family. It may not be wise to e-mail the whole image collection to every person you know, because e-mail boxes are known to become full every now and then. Instead, post your photos on an online photo album. These photo-sharing albums on the Internet, such as Buzznet, Flickr, Kodak Gallery, and Webshots, offer free accounts and an amount of free storage space for your photos.

caution *Before posting a single picture on the Internet, keep in mind that anything posted online may begin a life of its own. A funny picture of a friend posted on your album may pop-up years later in another context when, for example, your friend is getting ready for his career-defining job interview.*

Let's begin by figuring out the choices. All photo-sharing services allow registered users to post pictures from their computers. Many, but not all services, also accept images uploaded via e-mail or by a dedicated mobile application. On the other hand, there are things that your camera phone can do and things it can't do. For making photo sharing as easy as possible for you, these two—the photo-sharing service and your phone's capabilities—have to be matched.

Step 1: Plan the Upload Process

Make yourself familiar with the Internet access features on your phone, because you will need those features in this project. E-mail or MMS capability gives you access to a number of photo sharing services, but you may also be able to install a dedicated application that does it all for you.

1. Choose the Technique for Uploading Your Photos

The most flexible method for uploading your photos, which also gives you the most control, is e-mail. If you can send e-mail messages with attachments from your phone, you will be able to post to many photo-sharing sites on the Internet.

Not all phones come with e-mail, but practically all camera phones come with MMS (Multimedia Messaging Service) capability. MMS was originally designed for sending multimedia, such as pictures, voice, and video clips from one phone to another. However, it is possible to send MMS messages to e-mail addresses as well. This means that you can post photos via MMS to a photo-sharing service without using e-mail on your phone at all.

note *By design, MMS has some restrictions. Network operators who relay MMS messages between phones often restrict the maximum size of a message to 50–100KB. The photos you send by MMS should be about 640×480 pixels in size to go through the system. Although pictures of this size look fine on a small screen of a phone, they don't always look good when viewed on a computer monitor.*

In addition to e-mail and MMS, there's one more option for uploading photos. You can install a dedicated photo upload application on your phone. Products are available for Palm OS, Symbian OS, and Windows Mobile smartphones as well as for phones that can run Java software. Check if your phone is compatible with, for example, these products: FotoJive (www.fotojive.com), Shozu (www.shozu.com), SplashBlog (www.splashblog.com), and Yahoo Go (go.connect.yahoo.com).

2. Choose the Photo-Sharing Service

Many popular photo-sharing services can receive pictures from camera phones. If you already are a member of an online service, view its Web pages for instructions on how to use a camera phone for posting images.

If you want to sign up to a new service, you might consider one of these: Buzznet (www.buzznet.com), Flickr (www.flickr.com), Kodak Gallery (www.kodakgallery.com), PhotoBox (www.photobox.co.uk), Photosite (www.photosite.com), PictureTrail (www.picturetrail.com), or Webshots (www.webshots.com).

3. Choose Your Upload Process

Now, you are ready to decide on the photo-sharing process that works best for you. If you intend to share pictures every now and then, you might want to choose e-mail for posting. If you foresee that you will frequently share images, and you have a data communication plan that lets you use plenty of megabytes, an upload application is the best solution for you.

Step 2: Share Photos via E-mail or MMS

In this step, I will show you how to use e-mail for posting pictures on Buzznet photo album. Many other photo-sharing sites accept images sent via e-mail as well.

1. Make Sure You Have a Working E-mail Account on Your Phone

If you have successfully sent e-mail messages (with attachments) from your phone, you are all set for this project. See Project 10 to set up and troubleshoot e-mail access on your phone.

tip *Even if your camera phone doesn't have e-mail access, you can post to online albums via MMS. Ensure MMS works on your phone before proceeding with this project. Your service provider can give you the correct MMS settings if you have connectivity problems. Just remember that your service provider may charge extra for MMS messages.*

2. Write Down the E-mail Address Buzznet Has Reserved for Your Uploads

If you haven't registered to a photo-sharing service yet, now it is time to do so.

Log in to your Buzznet account at www.buzznet.com on your computer. Go to your Buzznet home page and click Upload Photos. Click Settings and scroll down until you find the address for posting to your gallery through e-mail. This is the e-mail address where you will send your photos.

Non-Buzznet users should look for the upload e-mail address from the personal profile page or from the account settings of the service.

3. Add the E-mail Address to Your Phone Contacts

You will need the e-mail address for posting photos many times in the future, so it's worth saving it in your phone's contact list.

Symbian OS/S60 Phone Go to the main menu and select Contacts | Options | New Contact. Type **Buzznet** (or whatever you believe describes your shared photo album) into the First Name field. Scroll down until you find E-mail (see Figure 12-1). Enter the address that was reserved for you in the Buzznet Settings.

Windows Mobile Phone In the Home screen, push contacts. Open the menu and select New Contact. Type, for instance, **Buzznet** into the First Name field. Scroll down until you find E-mail (see Figure 12-2). Enter the address that was reserved for you in the Buzznet Settings.

Figure 12-1

Save the e-mail address for your Buzznet photo uploads on your Symbian OS/S60 phone.

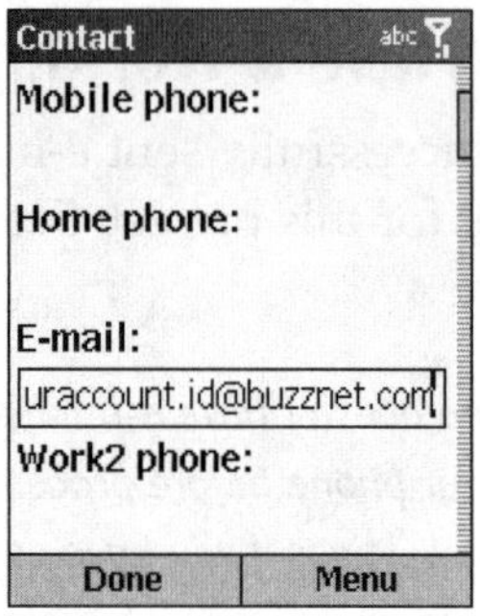

Figure 12-2

Add the Buzznet e-mail address into the Contacts on your Windows Mobile phone.

Other Phones Add the photo upload e-mail address to the phone's contacts or address book.

4. Send a Photo

Symbian OS/S60 Phone If you have just snapped a photo and want to post it right away, you can do it while the camera is still active. Select Options | Send | Via E-mail (or Via MMS/multimedia, if e-mail doesn't work on your phone).

Alternatively, you can send a picture you have taken earlier. In the main menu, open Gallery (or open the Media folder and launch the Gallery application from there). Select Images and find the picture you want to send. Select Options | Send | Via E-mail (or Via MMS/multimedia).

You should see a form for composing a message (see Figure 12-3). Select Options | Add Recipient. Pick up the address you saved for your Buzznet uploads. Select Options | Send.

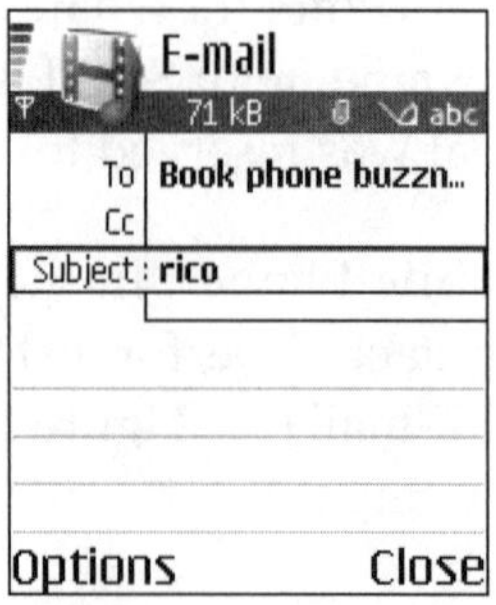

Figure 12-3

The number of kilobytes (KB) and the paper clip indicate that an attachment is included in the message.

Windows Mobile Phone Go to the Home screen and push START. Open Pictures & Videos (or Photo Album) and find the photo you want to send. Highlight the picture and open the menu. Choose Send and select your e-mail account (or MMS/Multimedia if you are not using e-mail) from the list. You should see a screen for composing a message (see Figure 12-4).

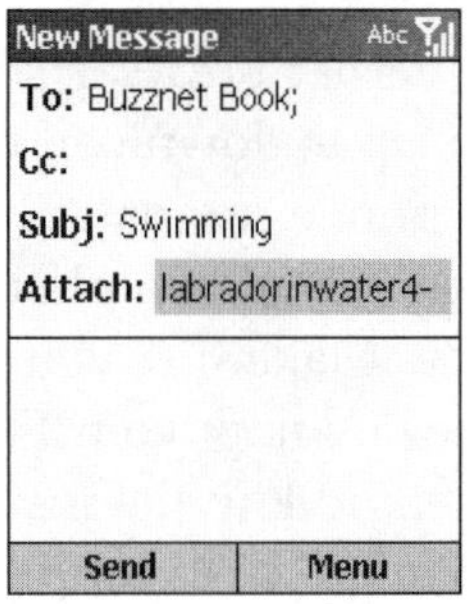

Figure 12-4
Pick up the recipient (in this case, your Buzznet photo album) and send the message.

When the To: field is active, open the menu and select Add Recipient (Insert Contact). You'll see a list of contacts. Pick up the e-mail address that you created earlier for photo uploads. Press SEND. The message will be sent the next time you access e-mail. For instantly sending the picture, go to the Home screen and activate Messaging. Open your e-mail account, open the menu, and choose Send/Receive.

note *If you don't have a working e-mail solution on your phone, try MMS. Usually, service providers have set limitations for the MMS attachment size. Phones powered by the Windows Mobile 5 software can automatically scale down the image size. Devices that are running on the older versions of the software, such as the Windows Mobile 2003, don't automatically scale down images; you have to do it yourself before sending the MMS.*

Other Phones Open an application that lets you view pictures. Find the picture you want to share and send it via e-mail or MMS.

tip *You can use MMS for posting photos for all photo sharing services that accept new pictures by e-mail. To avoid MMS communication problems, take your pictures in 640×480 pixel resolution, or scale them down to smaller than 100 KB.*

5. View Your Shared Photo Album

When you get access to a computer, take a look at the pictures on your online album. This is what your friends and family will see once you send them the link to the album.

Step 3: Share Pictures Using a Photo Upload Application

I am going to use the ShoZu application for uploading pictures to Flickr photo album in this step. With ShoZu, the setup and image upload process is the same for all photo-sharing sites it can connect to.

1. Install the Shozu Application on Your Phone

Launch the Internet browser software on your computer. Go to www.shozu.com and click Sign Up to download the application and to join the service. If you haven't checked your phone compatibility with ShoZu yet, do it before registration. Also, if you haven't signed up to a photo-sharing service (such as Flickr), do it before registering to ShoZu. Make sure you can log in to your photo album account because you have to let ShoZu know your Flickr (or other photo album account) user name and password. Then, follow the instructions on the ShoZu Web page to download (see Figure 12-5) and install the application.

Figure 12-5

During the sign-up process, you will receive a text message from Shozu. Click the link to start the application download (the picture is from a Symbian OS/S60 device).

Text message
1/37
From +4477654(
To install ShoZu select link
http://www.shozu.com/d/s60v8/shozuen.wml
Options Back

Before you leave ShoZu's Web pages, check that your Sharing settings are correct. You can also find out about features for sending, describing, tagging photos, and downloading video clips and music.

2. Activate ShoZu

When the ShoZu has been successfully installed on your phone, you have to activate it.

Symbian OS/S60 Phone Go to the main menu and open the My Apps folder (or My Own, Fun, or Install) to locate and launch ShoZu. It will ask for you to log in. Now, be careful: you must enter your ShoZu user name and password (see Figure 12-6), not your Flickr user name.

Figure 12-6

Activate the application on your Symbian OS/S60 phone by entering your ShoZu user name and password.

ShoZu Transfers
ShoZu username:
Abc
ShoZu password:
Cancel

Windows Mobile Phone In the Home screen, press START and find ShoZu in the menus. Launch ShoZu. It will ask you to enter your user name. Type your ShoZu user name and password (see Figure 12-7), not your Flickr login name.

ShoZu
ShoZu username:
book
ShoZu password:
Done
Cancel

Figure 12-7
Activate the connection to the photo-sharing service on your Windows Mobile phone by entering your ShoZu user name.

Other Phones Find ShoZu from your phone menu and launch it. Enter your Shozu user name and password for activating the application.

3. Post a Photo on Your Online Album

Now, everything is ready for you to start sharing your photos. Take a picture on your camera phone just as you would normally do.

Symbian OS/S60 Phone After the new image is saved on the phone or on the memory card, ShoZu wakes up (see Figure 12-8) and asks for confirmation for uploading the photo to Flickr. You can confirm or reject the upload or add details to the photo, such as title and description.

Figure 12-8
Let ShoZu upload the picture to your online photo album.

Windows Mobile Phone When the camera has saved the captured image in the memory, ShoZu wakes up. It asks for permission to upload the photo to Flickr. Confirm or reject the upload, or add details to the photo, such as title and description (see Figure 12-9).

Other Phones After you have taken a new picture, ShoZu wakes up. Confirm or reject the request to upload your photo to the Internet.

Figure 12-9

Confirm the upload and your photo will be automatically posted on your photo album.

note *You don't have to upload your pictures right away. Whenever you want ShoZu to upload photos, launch the application, select any saved photos, and let ShoZu upload them.*

tip *ShoZu usually realizes when you are outside your home network and informs you about potential roaming costs. In order to avoid potentially high international connection fees, you can tell ShoZu to stop uploading pictures while you are visiting overseas.*

4. View Your Shared Photo Album

When you get access to a computer, log on to your Flickr photo album, and take a look at your picture collection. You can view photos, organize photos, edit descriptions, or remove your photos, not to mention browsing millions of other people's pictures. Don't forget to let your friends and family know the link to your photo album.

Project 13

Upload Your Camera Phone Movies to a Video-Sharing Site

What You'll Need:

- **Account on a video sharing service on the Internet**
- **E-mail access on the phone**
- **Data communication (wireless Internet access) plan from your service provider**
- **Cost: The cost of sending photos to the Internet from your phone depends on your data communication plan**

Practically all camera phones that can capture photographs can record video clips. It is fun to watch the recorded clips on the phone screen, but it is even more fun to share them with family and friends. No matter how entertaining the clips may be, you shouldn't just e-mail your videos to every person you know. Even a short video clip tends to make up a large file that quickly fills up e-mail boxes. There is a better way to show your movies: upload them on a video-sharing service on the Internet.

You can easily find a number of video-sharing sites on the Internet, including Blip TV (www.blip.tv), Dailymotion (www.dailymotion.com), Revver (www.revver.com), vSocial (www.vsocial.com), and YouTube (www.youtube.com).

note *Sharing photos on the Internet is one thing, but sharing video clips is another thing entirely, because of the following:*

Even a short video recording can become a large file. The size of the file depends on the image resolution and how much the video software compresses images. Use the highest video resolution for sharing on the Internet and the lowest resolution for sending clips from phone to phone.

Double-check your mobile data communication plan. If you intend to frequently upload videos directly from your phone, you should consider a flat-rate plan that allows unlimited data traffic, or at least, a large number of megabytes of traffic per month.

Many video-sharing sites allow posting of new movies only from Internet browser software. All Internet browser products on a computer can upload files, but it still is rare for an Internet browser on a phone to have this capability. You can always copy the movie from your phone to your computer and upload it from there.

Some video-sharing services can receive clips via e-mail or via MMS. You can transmit your movies directly from your camera phone to these sites, for example, to blip.tv and YouTube.

Video-sharing services may require you to upload your video in a certain format, such as MOV (Quicktime), WMV (Windows Media Video), or MPEG4.

Despite the technical requirements, it is not difficult to share movies on the Internet once you have made all the necessary preparations. The really nice thing is that the best video-sharing services do some of the work for you. For instance, they convert a myriad of video formats into a commonly used format that can be watched on any computer without having to install any additional software.

In order to make sure that you find a technique that works both for your phone and for your video-sharing site, we'll go through three cases: upload by e-mail, upload from a computer, and upload from a phone browser software. Pick the one that suits you the best.

E-mail is the easiest option to use for uploading videos. If you can't use e-mail, some smartphones come with an advanced browser software that can upload files, including videos, from the phone. This feature lets you post movies to practically any video-sharing site directly from your phone.

If you want to avoid data transmission charges altogether, or you'd like to minimize the use of megabytes from your data subscription plan, copy the videos to your computer and upload the clips to the Internet from there.

Step 1: Set Up Your Camera for Video Sharing

Before sharing any videos or even before shooting them, you should check the video settings on your camera phone. Most devices come with a low-quality video option for sending clips by MMS from one phone to another (see Figure 13-1). Usually, megapixel camera phones also feature a higher-quality video recording option (see Figure 13-2).

Memory space is also a consideration. For example, a 50-second clip captured in 176×144 (QCIF) resolution in 3GP format results approximately in a 500KB file. Digital cameras can capture video in higher resolution, and they typically don't compress images as heavily as camera phones do. For instance, a 320×240 (QVGA) resolution video clip about 60 seconds long and saved in AVI format takes about 10MB of memory space in a digital camera.

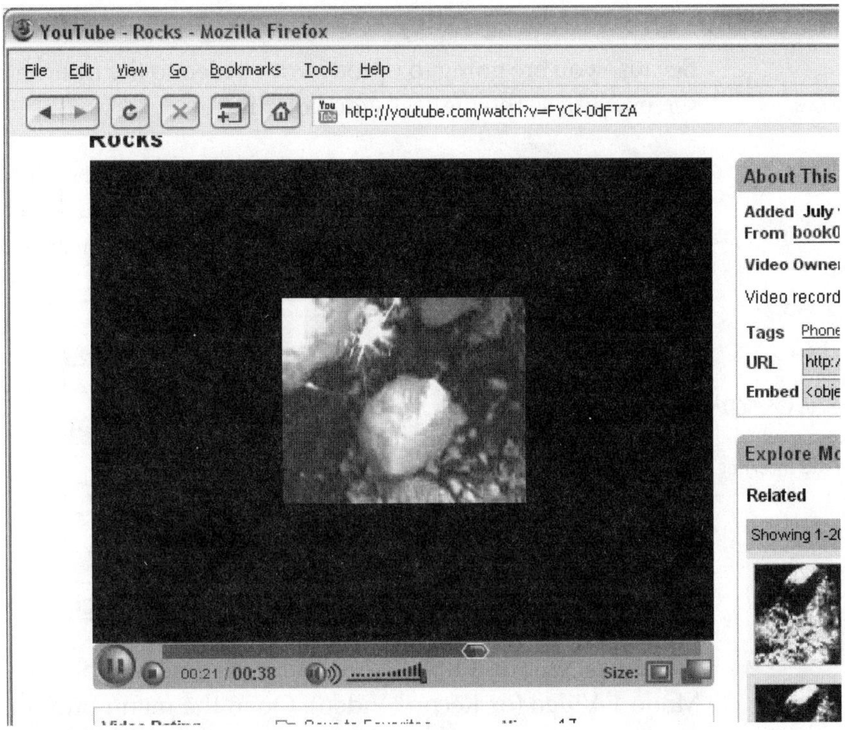

Figure 13-1

A video recorded in 176x144 resolution is suitable for viewing on a phone screen, but is small and grainy when viewed on a computer monitor (as seen here).

Figure 13-2

A video recorded in 352x288 resolution looks fine even on a computer screen, although the still image captured from the live video may not make justice to the quality.

1. Define the Video Quality

Because you are going to upload your video to the Internet for viewing on a computer screen, choose the highest video quality.

Symbian OS/S60 Phone Activate the camera. Open Options and select Video mode (or New and Video Clip). Select Options | Settings | Video Quality (Video Resolution) and choose the highest value (see Figure 13-3). Press BACK when you are done.

Figure 13-3

Select the best image quality on your Symbian OS/S60 phone for sharing videos on the Internet.

Windows Mobile Phone Activate the camera. Open the menu, and select Capture Mode | Video (or Record Video). Open the menu once more and choose Options. If you see a list of options, select Modes. Now, you can change the video settings (see Figure 13-4). Select the highest resolution. You can also change the Encoder to AVI or MPEG4 for the best image quality. Some phone models only let you select the Record quality. In this case, pick up High. Press DONE when you are ready.

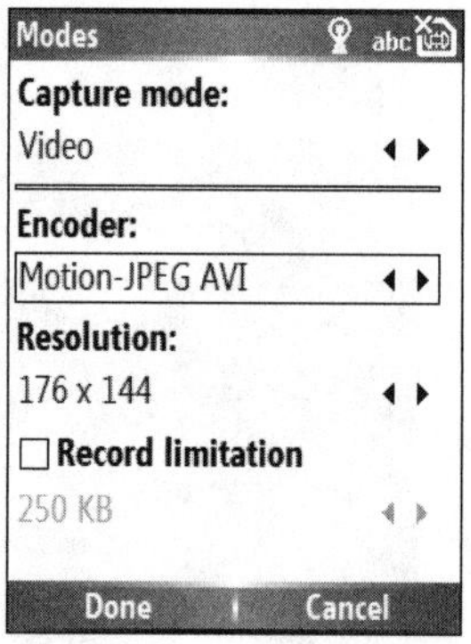

Figure 13-4

Select the best image quality on your Windows Mobile phone for videos that you intend to share on the Internet.

Other Phones Activate the camera and go to the video settings. Set the highest resolution and quality you can find.

tip *There is an exception to specifying the video quality. When you have to send your video via MMS, choose the lowest video quality, such as the highly compressed 3GP in 176×144 resolution. While you are recording a video that will be transmitted as an MMS message, try to keep the clip length under 10 seconds. This ensures the clip stays smaller than 100KB, which is often the limit for messages that service providers let through their MMS systems.*

2. Save Your Video Clips on a Memory Card

Before you record new video clips on your camera phone, make sure it is saving images on the memory card. You can change this setting by activating the camera, opening the menu, finding settings, and changing the default storage device to memory card.

Step 2: Share Your Videos via E-mail

In this project, I will show you how to use e-mail for posting videos on blip.tv (its Internet address is the same: blip.tv). There are other video-sharing sites that accept videos via e-mail, and the e-mail upload process is pretty much the same for all of them. For example, YouTube can also receive videos via MMS or from the Shozu application.

1. Make Sure You Have a Working E-mail Account on Your Phone

If you have successfully sent e-mail messages (with attachments) from your phone, you are all set for this project. Take a look at Project 10 if you need help setting up and troubleshooting e-mail access on your phone.

2. Write Down the E-mail Address blip.tv Has Reserved for You

If you haven't registered with a video-sharing service yet, now it is time to do so.

Log on to your blip.tv account on your computer. Go to your blip.tv home page and click Publish | Upload | Mobile Upload. Type a code word that will become part of your upload e-mail address. Then, you will get the full address where you can send your videos. Write it down.

If you use some other video-sharing site than blip.tv, you should look for the e-mail address from your profile or account settings.

3. Add the E-mail Address to Your Phone Contacts

You will need the e-mail address many times when uploading videos, so it's worth saving on the phone.

Symbian OS/S60 Phone Go to the main menu and select Contacts | Options | New Contact. Type blip.tv (or whatever you believe describes your video album) into the First Name field. Scroll down until you find E-mail. Enter the address blip.tv reserved for you.

Windows Mobile Phone In the Home screen, press CONTACTS. Open the menu and select New Contact. Type, for instance, **blip.tv** into the First Name field. Scroll down until you find E-mail. Enter the address that was reserved for you by blip.tv.

Other Phones Add the upload e-mail address to the phone's contacts or address book.

4. Send the Video

After you have recorded your movie, and perhaps watched it to make sure it qualifies, you are ready to post the clip on your online video album.

caution *Before sharing your home movies on the Internet, keep in mind that anything posted on a public web page may begin a life of its own. A funny video of a friend posted on your album may pop-up years later in another context when, for example, your friend's children are surfing on the Internet.*

Symbian OS/S60 Phone If you want to post a video right after you have recorded it, you can do it while the camera is still active. Select Options | Send | Via E-mail (or Via MMS/multimedia, if you are posting to a service that only accepts MMS messages).

Alternatively, you can send a video you have recorded earlier. In the main menu, open Gallery (or open the Media folder and then, Gallery). Select Video Clips and find the movie you want to send. Select Options | Send | Via E-mail (or Via MMS/multimedia).

You should see a form for composing a message (see Figure 13-5). Open Options and select Add Recipient. Pick up the address you saved in Contacts for your video uploads. Open Options and choose Send.

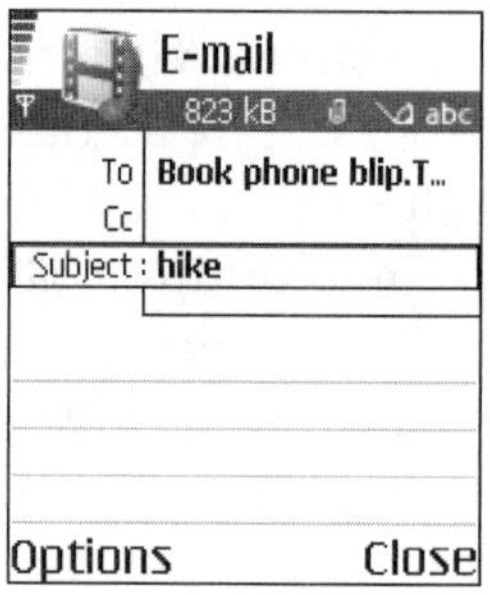

Figure 13-5

The number of kilobytes (KB) and the paper clip indicate that an attachment is included in the message.

Windows Mobile Phone Go to the Home screen and press START. Open Pictures & Videos (or Video Player) and find the video you want to send. Highlight it and open the menu. Choose Send and select your e-mail account (or MMS/Multimedia if you are not using e-mail) from the list. You should see a screen for composing a message (see Figure 13-6).

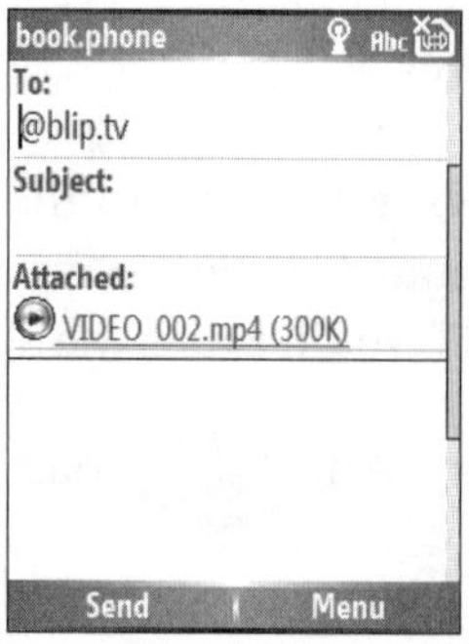

Figure 13-6

Pick up the recipient (in this case, blip.tv) and send the message.

When the pointer is in the To: field, open the menu and select Add Recipient (Insert Contact). You'll see a list of contacts. Pick up the e-mail address that you created for uploads. Press SEND. The message will be sent the next time you access e-mail. For instantly sending the video, go to the Home screen and activate Messaging. Open your e-mail account, open the menu, and choose Send/Receive.

Other Phones Open the photo/video viewer on the phone, find the video you want to share, and send it via e-mail or MMS to a video-sharing site.

5. View Your Shared Video Album

When you get access to a computer, take a look at the videos you have uploaded. This is what your friends and family will see once you send them the link to the album.

Step 3: Share Movies by Uploading Them from Your Computer

You may want to use your computer for uploading videos in any of the following cases: the recorded video is megabytes in size, you want to avoid mobile network costs, or the video-sharing site accepts uploads from computers only. The upload is done in two phases. First, you copy your video from the camera phone to your computer. Second, you upload the clip from your computer to your video album. Although I'm using YouTube as an example in this project, this particular method works with practically all video-sharing sites.

1. Remove the Memory Card from Your Phone and Insert It into a Memory Card Reader

Assuming you have saved your movie clip on a memory card, pull the card out from the phone, and stick it into a card reader. Copy the video clip from the memory card to your computer's hard disk. Make a note where you saved it. You may view a detailed description of this procedure in Project 2.

If there's no memory card on your phone, use Bluetooth or a data cable for transferring the video to your computer. Read Project 3 for details for using Bluetooth.

2. Edit the Video (If You Want)

If you want to improve your video's attractiveness, edit it in a video software application, such as Windows Movie Maker, Apple iMovie, or Adobe Photoshop Album. Video editing products allow you to keep the best segments of the movie and scrap the rest. Read Project 20 for instructions and tools for editing your videos. For example, YouTube recommends MPEG4 video format at 320×240 resolution with MP3 audio for best viewing.

Don't worry. You don't have to do anything to your video before uploading it to YouTube. The service will do the necessary format conversions. Typical camera phone video recordings, such as 3GP video in 176×144 resolution, AVI video in 320×240 resolution, and MPEG4 clip in 352×288 resolution all play fine in YouTube.

3. Log on to Your Video Album Account

Log on to YouTube at www.youtube.com from your computer. If you haven't registered with YouTube, or any other video-sharing service yet, do it now.

4. Upload Your Video

Once you have logged on to YouTube, click Upload. Fill in the required forms, locate the video file from your computer's hard disk, and click upload. Then, simply wait for your video to be copied to the server.

5. See How Your Video Plays on YouTube

To view your video album, click your account name at the top of the YouTube window. YouTube will list your videos. Click the thumbnail picture of the video that you want to play.

If the video looks more blurry or mosaic than it was on your phone screen, click the small box with a square around it on the lower right-hand corner of the video window. The video will be rescaled to its original size, which will make it look sharper on the large screen.

Don't to forget to send the link to the video album to your friends and family.

Step 4: Share Videos by Uploading them from Your Phone's Internet Browser

It is easy to upload videos or photographs using an Internet browser on a computer. Browser software products on phones, however, don't typically come with the required feature for uploading files. There are some smartphones, like the Nokia E- and N-series, that include a browser that is capable of uploading files. Also, the Opera browser software can upload files from the phone.

A definitive way to find out if the browser on your phone can upload files is to try it yourself. Launch the Internet browser on your phone, log on to your video album account, and go to the upload section. If you can't see the upload button or can't pick up the video file from the phone (see Figure 13-7), you won't be able to use this technique.

If your phone browser application can upload files, the rest is easy—use it like your computer browser.

Figure 13-7

The browser software (Opera in this picture) must be capable of picking up a file from the phone memory or memory card for uploading it to the Internet.

1. Log on to Your Video Account

Launch the browser software on your phone and go to your online video album. Log on to your account.

2. Upload a Video

Navigate to the Upload or Posting section. Fill in the required information and click the button that lets you select the file to upload. Pick up the video from the phone memory or memory card and click the Upload or Post button.

3. Watch Your Videos

You still have to go back to your computer to watch movies posted on video-sharing sites. Phones can play movies, too (you just recorded and viewed one). But when you upload your video to the video-sharing service, it may convert the clip to another format that your phone doesn't recognize. For example, watching videos on YouTube requires Macromedia Flash software. Most computers have the Flash software installed, but very few mobile devices come with it.

tip *The Shozu photo upload application—which is discussed in more detail in Project 12—can upload videos as well. If you have, for example, a YouTube account, you can shoot a video on your camera phone, and when you save it, Shozu can automatically upload it to YouTube.*

Project 14

Record Your Own Ringtones

What You'll Need:

- **A voice recorder application on your phone**
- **A ringtone editor application for your computer**
- **Cost: $20 for Ringtone Media Studio**

Do you remember the rather annoying ringtone where someone was imitating the sound of a speeding motorcycle? If nothing else, it proved that practically anything can become a hugely popular ringtone. Creating a tune like that isn't difficult. You simply record sounds that you can find anywhere around you (see Figure 14-1). In this project, we are going to capture audio clips on a phone or on a computer and turn them into ringtones.

If your phone doesn't include a voice recorder application, or you want to edit the captured audio clip, mix your new ringtone on a PC. For example, Windows Sound Recorder or Ringtone Media Studio are easy-to-use applications for turning self-recorded audio clips into ringtones.

Let's begin by creating a new ringtone on the phone. If this is not possible on your device, skip to the section where recording and mixing new ringtones on a PC is discussed.

c|net See a CNET video on recording your own ringtones at http://diyphone.cnet.com

Step 1: Create a New Ringtone on Your Phone

In this step, you'll learn to record any audio (or noise) sources around you and turn them into ringtones.

Figure 14-1

Let your imagination fly. An endless resource of free ringtones may be closer than you think.

1. Record an Audio Clip

tip *Place the phone microphone close to the audio source before you start recording. The microphone is practically always located at the lower end of the device. Often, it is only a tiny hole in the body of the device.*

Symbian OS/S60 Phone Go to the main Menu and open the Media folder (or Office folder). Launch the Recorder application.

note *Just in case you want to transfer the recorded clips to your computer for further mixing, define the default storage device in the Recorder application. Open the Options menu and select Settings. If Memory Card is not selected as the Memory in Use, press selection key to change the value to Memory Card.*

In the main screen of the Recorder application, highlight the red record button. Press the selection key, and the recording will start (see Figure 14-2). Press STOP to end it.

Figure 14-2

When you record, keep it short—10–30 seconds should be enough for a good ringtone.

Voice recorder
Sound clip(01)
27/07/2006 - 16:19
Length: 00:14
Recording
0 01...
Pause Stop

Windows Mobile Phone Go to the Home screen and push start. Locate the Voice Notes (Voice Recorder) application and open it. You may see an empty screen, but it only means you haven't recorded anything yet.

When you are so close to the audio source that the phone microphone can pick up the sound, push RECORD (see Figure 14-3). End the recording by pressing STOP.

Although you can record as long as there is memory space available, try to limit the recording to 10–30 seconds. Windows Mobile phones record audio in WAV format

Figure 14-3

Record anything that the phone microphone can pick up.

which tends to result in large files. A long recording will take plenty of memory space, and your phone may not be able to play a lengthy tune anyhow.

Other Phones Open the recorder application and record an audio clip.

2. Name the Tune

If you develop an ear for catchy sounds, you may soon have a collection of fabulous ringtones on your phone. Give the tunes descriptive names so that you can identify them later. Who knows, perhaps your new ringtones are so good that your friends will want to have the best ones as well.

Symbian OS/S60 Phone In the Recorder application, select Options | Rename Sound Clip. Type a new name for the tune. You can change the name of the last recording only.

Windows Mobile Phone When you are in the main screen of the Recorder, highlight a recording in the list. Open the menu and select Rename. Type a new name for the tune.

Other Phones Open the menu in the audio recorder application. Look for a menu entry that lets you rename the saved recording.

3. Set the Recording as Your New Ringtone

Symbian OS/S60 Phone With the Voice Recorder is on the screen, select Options | Go to Gallery. All the recordings you have saved can be accessed from the Gallery as well, without having to use the Recorder. You should see a list of recorded tunes (see Figure 14-4). Highlight a recording, and select Options | Set as Ringing Tone.

Windows Mobile Phone When you are still in the Voice Recorder, highlight a recording in the list and open the menu. Select Set as Ringtone (see Figure 14-5).

If the Recorder menu on your phone doesn't show the Set as Ringtone choice at all, you have to set it manually. Go to the Home screen, push START, and open File Manager. You have to move the recording from the folder /Storage/My Documents/Notes to the folder /Storage/Application Data/Sounds. Then, go to the Start menu and select Settings | Select Sounds | Ring Tone and pick up the recorded ringtone.

Figure 14-4

Your recordings on your Symbian OS/S60 phone are saved in the Gallery where you can select them as ringtones.

Figure 14-5

Activate the new ringtone on your Windows Mobile Phone.

Other Phones Open the menu of the voice recorder application and find a menu item that lets you set the file as your new ringtone. If you can't find it, go to the phone settings and open the ringtone settings for changing the new tune as the default ringtone.

Step 2: Mix a Ringtone on Your Computer

If you were lucky when you recorded the audio clip on the phone, you managed to catch a perfect sound that is suitable for a ringtone just as it is. Often, however, recordings contain unintended silence, cracks, and other noises that spoil the whole ringtone. Nonetheless, somewhere in the middle of the noise there may be a brilliant audio segment that would make the world's best ringtone.

You can save the best piece of your recording by transferring the clip to your computer and mixing a ringtone out of the rest of the noise. You need an audio recording/editing application on your PC for doing it.

If your phone came with a recorder application that can capture WAV files, you don't have to install any new software on your PC. For example, Windows Mobile smartphones record audio in WAV format. Windows PCs come with an application called Sound Recorder that can read, edit, and save WAV audio files. You don't have to install any additional applications.

On the other hand, phone owners whose voice recorder applications capture audio in mobile-specific format, such as AMR, need an additional software product for the PC. I'm going to use Ringtone Media Studio (www.ringtonemediastudio.com) as the audio mixer application in this project. Install the product on your PC if you want to be able to edit all types of ringtones.

tip *You can to use your computer for recording audio as well. Many notebook computers come with built-in microphones. You may also purchase a low-cost headset or microphone and hook it up to the microphone port in the computer. On a Windows PC, launch the Sound Recorder application (you can find it in Accessories | Entertainment) and push the red record button for saving an audio clip.*

Let's start from the moment when you have saved an audio clip on your phone but you want to modify it.

1. Copy the Recording to Your Computer

First, you need to transfer the audio clip to your computer's hard disk. Use a memory card (see Project 2 for details), Bluetooth (Project 3), a data cable, or e-mail (Project 10) for copying information from your phone to your computer.

2. Mix the Tune

Once you have successfully saved the audio recording on the hard disk, you can launch an application that lets you finetune a perfect ringtone out of it.

Sound Recorder The Sound Recorder is for WAV recordings only. You can find the application on your PC by clicking the Start button in the menu bar. Then, select Accessories | Entertainment | Sound Recorder.

Click File in the Sound Recorder menu and select Open. Find the WAV file from the hard disk and open it.

Select the Play button in the middle of the Sound Recorder screen to listen to the recording (see Figure 14-6). When you are listening, find the segment that you want to keep. Once you have identified the beginning of your new ringtone, stop the playback. Open the Edit menu and select Delete Before Current Position.

Figure 14-6
Save the best segment as your new ringtone.

Play the recording until you find the end of the new ringtone and stop there. Open the Edit menu and select Delete After Current Position.

Save the new audio clip. Open the File menu, select Save As, and type a new name for the tune.

If you have microphone installed on your computer, you can record new sounds on the computer as well.

Ringtone Media Studio Open Ringtone Media Studio. Click the My Melodies tab at the center of the application screen. Open the drop-down list (just below the tab) for finding the audio clip that you saved on the hard disk.

Figure 14-7

Select the best segment for your ringtone.

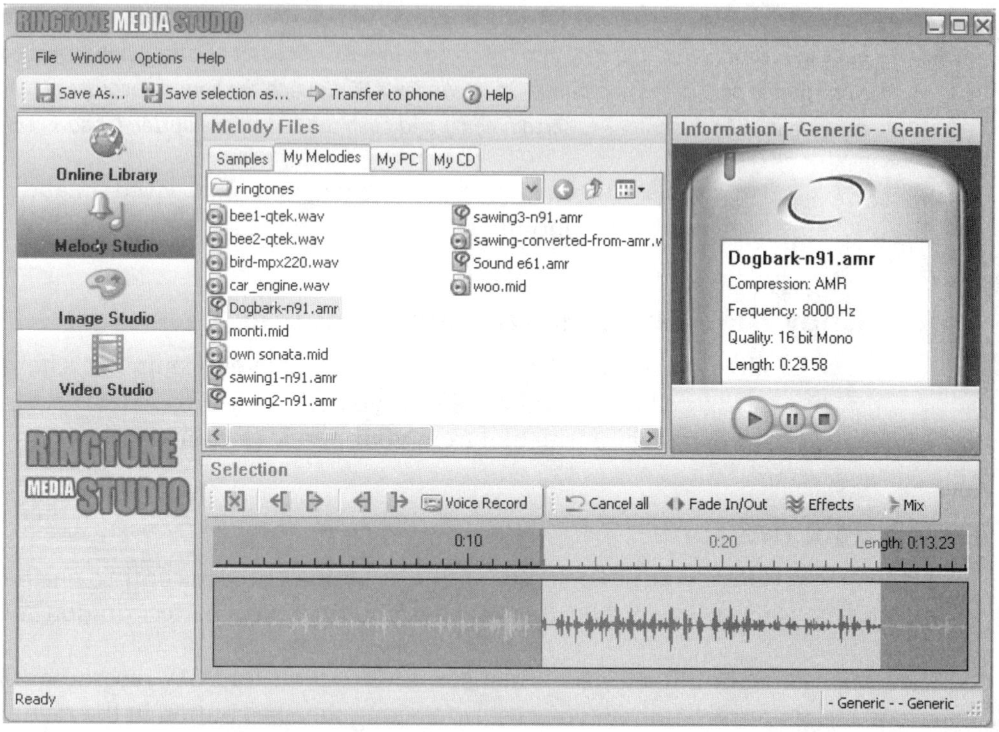

Play the clip by selecting the Play button in the pane on the right. Find the start point and end point of the tune you want to create. In the sound wave area at the bottom of the window, move your mouse pointer to the beginning of the new tune. Hold down the left mouse button and drag until you are at the end of the new tune. Release the mouse button (see Figure 14-7). You can test how it plays by clicking the Play button.

Once you are happy with the new tune, click Save Selection As at the top of the window. You can rename the clip and also save it in another ringtone format that's compatible with your phone. For example, saving the ringtone as a WAV file makes it compatible with most phones, but the file size increases.

Ringtone Media Studio includes a feature that lets you transfer ringtones directly from the application to your phone. Click the Transfer to Phone button at the top of the window and follow the instructions.

3. Transfer the Tune to Your Phone and Set It as Your New Ringtone

The new tone is ready and you are ready to let it ring to the world. Before that, you have to copy the file back to the phone.

Use a memory card, Bluetooth, a data cable, or e-mail to transfer the file to your phone. View Project 6 or Project 7 for advice on copying information from your computer to your phone.

Don't forget to set the tune as your new ringtone. Go to your phone settings and change the tune. You can find detailed instructions in Project 6.

Project 15

Display a Photo Slide Show on a Large Screen

What You'll Need:

- **A large screen. You can use your computer monitor, an external monitor, a TV, or a data projector**
- **A video cable**
- **Image viewing software. A photo album, image viewing, or photo editing application with a slide show feature will help you manage the slide show**
- **Cost: $5–$25 U.S. for a video cable. $300–$1000 U.S. for a large-screen monitor or TV**

One of the greatest advantages of digital photography over traditional film photography is the possibility of viewing photos on a large screen. Photos look more impressive and colorful, and some images even gain additional dimension of depth on a fluorescent screen, when compared to postcard-size print images. Before the digital era, photographers organized slide shows by reflecting photos onto the wall with a slide projector. The hassle with the projector setup, sorting dusty slides, furniture alignment, and light adjustments is history. We are going to set up a modern digital slide show in your living room in this project.

The Photo Album or Gallery application on your camera phone can display photos as a slide show on the phone screen. That's a fine feature for quickly viewing what actually has been saved in the phone memory. When you want to show your photos to other people, you want a large screen and you need to make some preparations.

It is rare for a camera phone to come with a connector that would let you directly hook it up to a TV or to a computer monitor. Nonetheless, you can organize a slide show that consists of your camera phone pictures. The options for displaying your photos on a large screen are:

- Transfer your photos to your home computer and store your pictures on its hard disk. Then, run the slide show in the image viewing software on your computer.
- Get an additional piece of hardware that connects your camera phone to a TV. In this setup, you can control the slide show from your phone.

Let's start by choosing the monitor and video connection for your computer. Let's also make sure you have a software application that allows you to organize the pictures and run the show from the comfort of your sofa.

Step 1: Choose the Screen for Your Slide Show

The larger the screen for the show is, the more impressive the images will be. That's true for good and poor images: great photos truly shine on a large screen, but poor photos—well, no one will know if you don't show your weaker shots at all. Depending on the equipment you have at home, your options for the slide show screen are a TV set, a computer monitor, or a data projector.

TV Set Displaying the photo show on a TV is often the preferred choice, because the TV set is already in the living room and the seating is arranged accordingly. Especially if you happen to own a fairly new wide-screen TV that has lots of interface options, the TV set could be the best choice (see Figure 15-1). Even if the photos don't look as sharp on your TV as on your computer monitor, it doesn't mean there is anything wrong with the TV. The resolution of computer monitors is usually higher than that of TV screens, which makes images appear more crisp and clear on computer monitors.

Figure 15-1
A notebook computer connected to a TV for watching a slide show.

Modern TV sets based on plasma or liquid crystal display (LCD) technologies are excellent choices for displaying photos. LCD and plasma monitors typically have high image resolution, and they don't flicker.

Computer Monitor A large computer monitor is another good choice for displaying a slide show. The advantages of a computer monitor are its high resolution (compared to that of TV) and easy connectivity.

If you can, use an LCD monitor (see Figure 15-2), as it doesn't flicker and the image is crisper than on a CRT monitor. However, LCD monitors have a limited viewing angle and you have to make sure before the show that your audience can see the screen. A high-quality cathode ray tube (CRT) monitor usually has the luminance and wide viewing angle to accommodate a larger audience.

Figure 15-2

A high-resolution LCD monitor displays sharp and clear images.

tip *A wide-screen notebook computer is a convenient choice for the slide show because of its portability and compact design. All you need for a photo show is already there. Just make sure the battery is fully charged before you unplug it and carry it to the stage. Place the laptop in a good position in the living room. For instance, set the laptop on a chair and move the chair to a location where it is close to viewers and no lights reflect on the screen.*

Data Projector Data projectors are excellent devices for displaying photo shows, because they have been designed to display high-resolution images transmitted from a computer. It is possible to display sharp images that are the size of a wall if the room is big enough and leaves space for the beam from the projector. The projector requires a darker room than a computer monitor or a TV. Data projectors cost between $1000 and $2000 US, but they can be used for watching videos and TV as well.

Step 2: Choose the Slide Show Software

In addition to a large screen and a computer, you need an application that manages the slide show. In theory, you could manually browse and display the photos one by one, for instance, using photo editing software, but that's too painful for the audience. To avoid boredom, it's vital to prepare the show well beforehand, and for that you need slide show software.

The application doesn't have to do much, but it has to do one thing well: it must automate the actual running of the show. Regardless of the software product used, the process of creating a new show is simple: collect the photos that make up your show, sort them into the order you want, and set the timing for how long each photo will be displayed. That's all you have to do, although many products let you add a variety of bells and whistles, such as background music and transitions between pictures.

If you don't already have image editing/photo album software installed on your computer, you can download free image viewing software, such as IrfanView (www.irfanview.com), Picasa (picasa.google.com), or XnView (www.xnview.com) (see Figure 15-3).

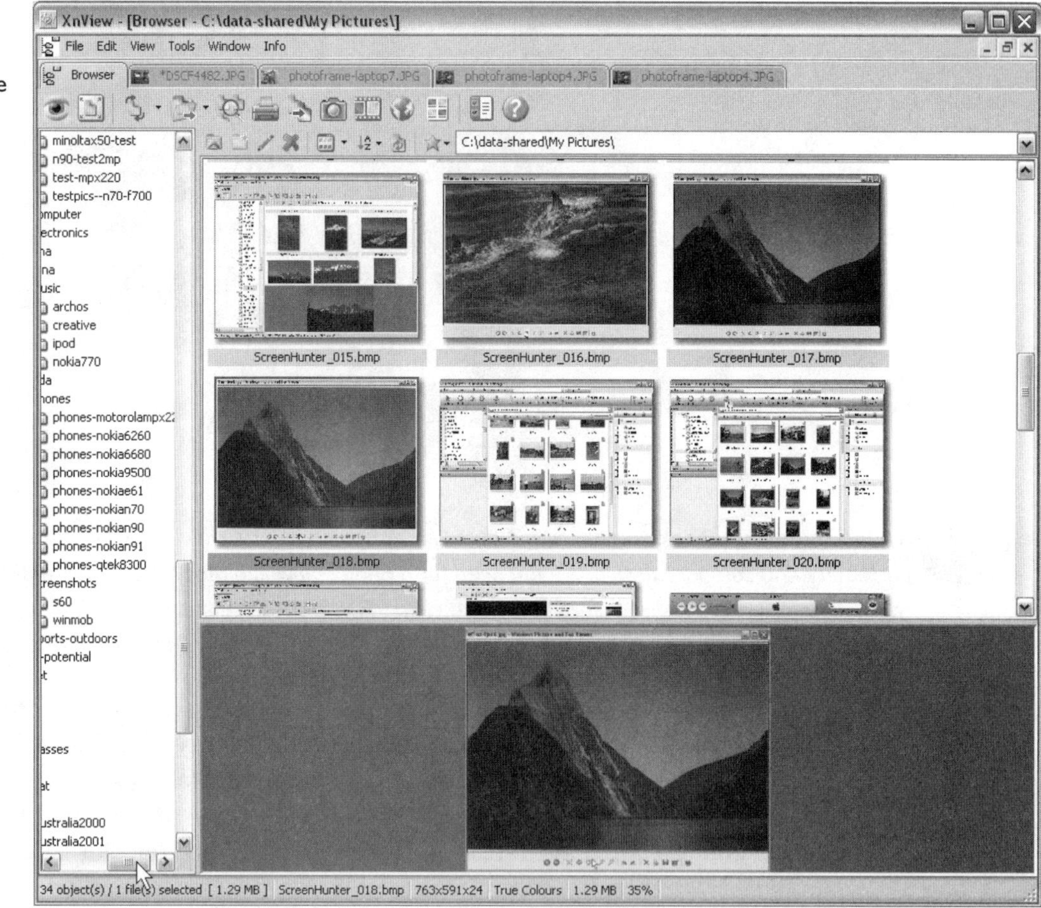

Figure 15-3

You can download free imaging software such as this XnView picture viewer application to manage your photo slide show.

Windows XP users can try the Windows Picture and Fax Viewer application as well. First, use the Windows Explorer application to collect the photos you intend to show in one folder. Then, right-click the first photo and select Preview. A new window should open and display the selected photo (see Figure 15-4). Click the Start Slide Show icon at the bottom of the window, and your slide show will begin.

Figure 15-4 Windows Picture Viewer lets you run slide shows, too.

You can do miracles to your photos if you take the time to learn an imaging application. You can sharpen blurry photos, lighten up murky mug shots, fade out unwanted objects, or apply artistic effects to the images. You need commercial image editing software to do this. Some image editing products, like ACDSee (see Figure 15-5), Adobe Photoshop Elements, and Corel Paint Shop Pro/Photo Album also include photo album and slide show features.

Step 3: Copy Your Photos from Your Phone to Your Computer

If you haven't transferred your photos from the camera phone to the computer yet, now is the time to do so. A memory card and Bluetooth are straight-forward techniques for transferring information between a phone and a computer. You can find detailed instructions for copying photos to the computer in Projects 2 and 3.

Step 4: Connect Your Computer to a TV or an External Monitor

If you have decided to use an external monitor for your slide show, you must find a cable that connects your computer to a monitor or TV.

Figure 15-5

This ACDSee photo management application includes image editing, photo album and slide show features.

S-Video S-Video is a common video interface in TV monitors and notebook computers (see Figure 15-6). The cable is also easy to connect. Purchase an S-Video cable from an electronics shop, and connect it from your computer to the TV set.

Figure 15-6

S-Video is a common video interface in notebook computers and TV sets.

When you invest in an S-Video cable, get as long a cable as you can. Then, you can comfortably sit on a couch and run the show with the notebook computer in your lap.

SCART or RCA Connection You may have an S-Video connector on your computer, but not on your monitor. If there is only an RCA connector (see Figure 15-7) or SCART interface (see Figure 15-8) on your TV, it is still easy to find a cable to hook up your computer to your TV. Purchase a cable with an S-Video plug on one end and a SCART or RCA plug on the other end.

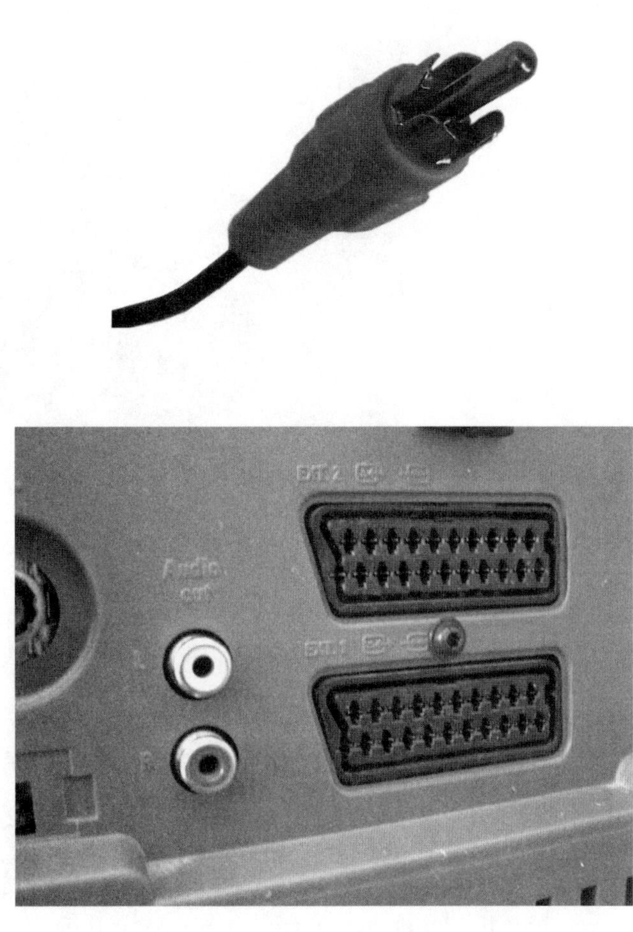

Figure 15-7

An RCA plug can be used either for video or audio.

Figure 15-8

SCART interface can be found at the back of TV monitors in some regions.

Graphics Card If there is no S-Video connector on your computer, one possible solution is to get an add-on graphics card. You can install a graphics card into a desktop PC, but not into a notebook computer. The graphics card must be installed into an expansion slot, but notebook computers don't come with these. Usually, the card can do many other things besides transmitting what is on the computer screen to the TV.

VGA-to-TV Adapter Yet another solution for computers without an S-Video interface is to purchase an adapter box that converts VGA signals from a computer to ordinary video signals that the TV can display. This solution works in notebook computers as well.

VGA Monitor A high-resolution computer monitor displays sharp images, and the connection is easy to make. Both LCD and CRT computer monitors come with a VGA cable that you can simply plug into the VGA interface on your computer (see Figure 15-9).

Data Projector Data projectors are easy to connect to a computer. Plug in the VGA cable from the projector to the computer's VGA port.

Figure 15-9
A VGA plug and port.

Step 5: Connect the Cable

Hook up the cable from your computer to the monitor, TV, or projector. Switch on the external monitor. Make sure the image viewer program on your computer is running.

Step 6: Change the Monitor's Video Input

TV Set Change the channel on your TV to "External" or "Video." There may be several channels for external sources on the TV, so you may have to test which one receives the video signal from your computer.

VGA Monitor Most VGA monitors can automatically sync to the video signal coming from a computer. Usually, you don't have to change the monitor settings in order to display images from a computer.

Data Projector Switch the signal input to VGA, or computer on the projector's control panel.

Step 7: Redirect Your Computer Video Signal to an External Monitor

Now you should have the cable between your computer and the TV or external monitor connected, and the monitor set for displaying the video signal from your computer. There's one more thing for you to do: you have to transmit the video signal from your computer to the external monitor. Depending on the type of connection you chose between the computer and the external monitor, the setup slightly varies.

S-Video Connected to TV Some S-Video connections are so easy that the picture appears on a TV without any further adjustments. For example, an old and trusty

Windows 98 notebook that I'm using for living room slide shows automatically transmits the video signal to the S-Video interface when the cable is connected.

If your TV doesn't display the screen from the computer (even when you have changed to the External or Video channel on the TV), you have to configure it manually.

Launch Control Panel on your computer. Select Display | Settings | Advanced | Monitor and you'll see the available monitors (see Figure 15-10). Click the red button on the TV tab and click Apply.

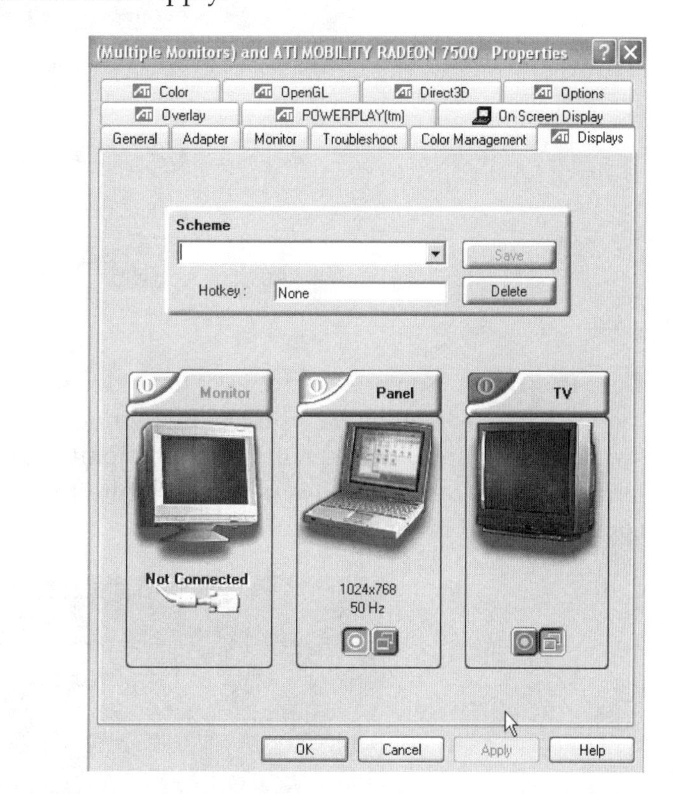

Figure 15-10

Activate the TV connection by clicking the red switch on the TV tab.

If you have a wide-screen notebook, there's one more thing to do. You must set the notebook screen in normal (non-wide-screen) mode. After you have clicked the red TV button, Apply, and then OK, you should be back in the Settings window (see Figure 15-11). Drag the Screen Resolution slider to 1024×768 (or 800×600, or whatever works for your TV), and click Apply.

VGA Monitor If you have connected an external VGA monitor to your notebook computer, you need to redirect the video signal to the monitor. Usually, there is a key combination that switches the display between the external monitor and the notebook's own screen. The exact key combination varies from computer to computer, but a fairly common one is FN-F7 (see Figure 15-12). Hold FN down and press F7.

Graphics Card Each add-on graphics card has its own configuration routine. You should see the included instructions for redirecting the computer screen to a TV.

VGA-to-TV Adapter View the adapter user guide for instructions on connecting the adapter to the TV.

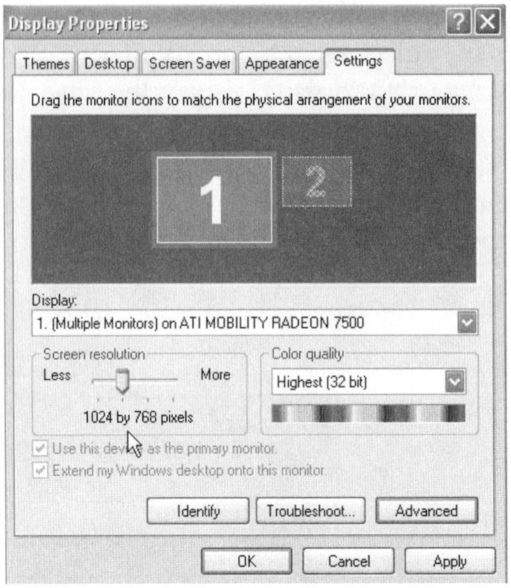

Figure 11

Reset your notebook's wide-screen mode to normal resolution for the TV connection.

Data Projector You need to redirect the video signal from a notebook computer to the VGA port where the data projector is connected to. Typically, a key combination switches the display between the external monitor and the notebook's own screen. A fairly common combination is FN-F7 (see Figure 15-12), but it varies from computer to computer. Hold fn down and press F7.

Figure 15-12

The key combination Fn-F7 lets you switch to the external monitor on this laptop computer.

If the computer screen doesn't show up on the TV or external monitor, you may have to restart your computer. Leave the external monitor on and the cable connected; power off your computer, and restart it. Then, re-try redirecting your computer's video signal to the external monitor.

Step 8: Display Your Slide Show

Launch the application you have chosen to manage the slide show. Make sure you know how to pause the show, skip a picture, and go back to the previous picture.

Select the photos that make up the show. Some applications let you add special effects and sync music with the show, but don't overdo the effects. Live narration and live questions and answers may be the best part of the show, after all.

If you have decided to run your slide show in the Irfanview application, select File | Slideshow from the program menu. Browse the folders on your hard disk until you find the pictures you want to display. Select images and click the Add button. When you have all the slide show pictures listed in the left pane—and you have sorted them into the desired order—you can start the show by clicking Play.

You can start a slideshow in Picasa by selecting View | Slideshow from the menu. Picasa will automatically display pictures from the folder you are currently viewing.

In Xnview, select Tools | Slide Show to create a slide show. Click Add to browse your photo folders. Once you find a folder or pictures your want to insert to your show, click Add. When all the images required for your show are listed in the File list–pane, click Go to start the show.

You Can Run the Slide Show from Your Phone

It is possible to avoid the use of a computer altogether when setting up a slide show system. You can connect your camera phone to an adapter device that controls the show displayed on your TV. This requires some extra equipment. For example, the following products let you make the required connections:

- An advanced memory card reader, such as the SanDisk Photo Album, or V-Mate. Save your camera phone photos on a memory card and insert the memory card into the SanDisk card reader. Connect the unit to a TV, and it will display the pictures on the TV.
- Nokia Image Viewer SU-5 is another device that reads images from a memory card, or from a camera phone via a data cable and displays the pictures on the TV.
- An adapter that can receive information via Bluetooth and transmit the data to the TV. Your task is to send images from your phone to the adapter. For example, the Anycom Bluetooth TV Adapter is a small pocket-size product that can do this.
- Sony Ericsson MMV-200 features both a memory card reader and Bluetooth. You insert your memory card or send your images from the camera phone to the unit over Bluetooth. Either a TV or a VGA monitor can be connected to the device for displaying pictures.

Some of these products can only display low-resolution photos. You may have to scale down your pictures to 640×480 or 800×600 pixels before sending them to an adapter. Follow the instructions included with the product to connect the cables, to setup the device, and to display your slide show.

Project 16

Print Photos from Your Camera Phone

What You'll Need:

- **Printer with USB port, Bluetooth connection, or memory card slot**
- **Your camera phone must have the matching (USB, Bluetooth, or memory card) feature in order to connect it to the printer**
- **A printing application on the phone. For example, many Symbian OS/S60 phones come with printing software, but you have to install an additional application on Windows Mobile smartphones**
- **Paper. Glossy photo paper makes all the difference when printing photos**
- **Color print cartridges for your printer**
- **Cost: $5–$30 U.S. for a printing application. The cost of glossy paper varies between $0.20–$1.00 U.S. per sheet. $10–$40 U.S. for a color ink cartridge. Also, read the note Be aware of the printing costs for additional information about total costs for printing photos at home**

One of the best things with digital photography is that you don't have to think twice about the costs when you go about snapping pictures. However, if you want to have a picture of your loved one on your desk, or you realize it is time to send grandma her grandchild's birthday photo, you need to print photos from your camera phone. Then, you have to reserve a small budget for printing.

You can copy pictures to your computer and print to your home printer, or you can print pictures directly from your camera phone. If you want to leave printing to a professional, visit your local photo lab and pass the shopkeeper the memory card from your phone. You don't have to leave home at all if you have uploaded your photos on a photo-sharing site on the Internet. Many of these photo album services deliver prints to your home.

note *Printing photos on a home computer gives you good control over the layout, size, frames, text, and other elements of the printout. It is a fairly straightforward task if you master the basics of using a computer. Here's what you have to do if you decide to use your computer for printing:*

1. *Copy your photos from the camera phone to the computer. Read the detailed instructions in Project 2 for using a memory card as the transfer medium or Project 3 for transmitting photos over Bluetooth.*

2. *Enhance your pictures, or print the images as they are. An image editor application lets you change the light, colors, size, and shapes, or you can add artistic effects on the pictures before printing them.*

In this project, we are going to print pictures directly from your phone to a home printer. To ensure that you can find at least one way to print, we'll go through three printing techniques: Bluetooth, memory card, and USB cable. Let's start with the easiest one: Bluetooth.

Step 1: Print via Bluetooth

A number of new printers, for instance, from HP, Epson, and Canon come with built-in Bluetooth or infrared wireless connectivity. If your phone has Bluetooth (or infrared) capability and you have a printing application on the phone, it is easy to print photos wirelessly.

1. Turn on Your Printer

Switch your printer on and activate Bluetooth. Make sure that there is paper in the tray and color cartridges are not empty.

2. Define a Printer for Your Phone

Install a print application on your phone if you can't find one in your device. You can log in, for example, to Handango (www.handango.com) and search for products compatible with your phone. Install the downloaded product according to the provided instructions. Then, you can setup your phone for printing.

Symbian OS/S60 Phone Go to the main menu and open the Gallery (or Media folder and Gallery) application. Select Images. Find the photo you want to print and open it. Open the Options menu and select Print (or Printing Options and Printers). Select Options | Settings | Default Printer | Bluetooth | OK (see Figure 16-1).

The phone will search for Bluetooth-capable printers (see Figure 16-2). Select your printer from the list.

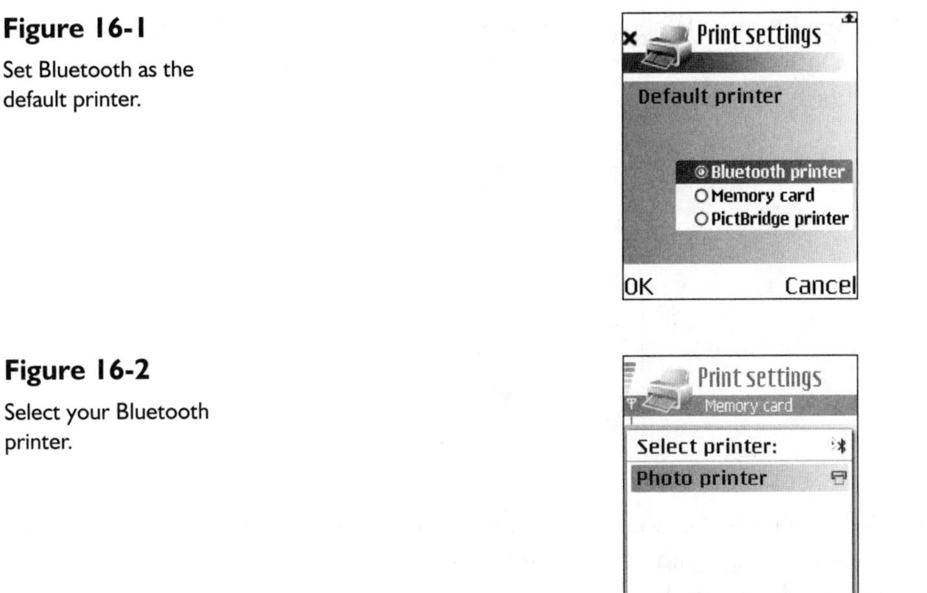

Figure 16-1
Set Bluetooth as the default printer.

Figure 16-2
Select your Bluetooth printer.

If the menu entry on your phone was labeled as Printing Options, you must add a new printer. Select Printing Options | Printers. In the Options menu, select Add. Go to the Bearer field and change the value to Bluetooth.

note *If you can't find Print at all in the phone menu, check once more the user guide and the software CD included with your product. The Print feature is actually an application of its own and may not be included in all S60 phones. You may also check out software download stores, such as Handango (www.handango.com) for products compatible with your phone.*

Windows Mobile Phone In the Start menu, find the printing application you have installed and open it. In this project, I used the Jetcet Print software (see Figure 16-3) for printing over infrared. The product is available for download at Westtek (www.westtek.com).

Figure 16-3
The Jetcet printing application for Windows Mobile phones.

3. Print

Switch on Bluetooth on your phone, or infrared if you are using Jetcet software on a Windows Mobile device.

Symbian OS/S60 Phone When you are viewing the photo you want to print in the Gallery, select Options | Print (or Printing Options). The printing will start to your default printer (see Figure 16-4).

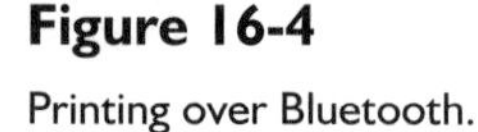

Figure 16-4

Printing over Bluetooth.

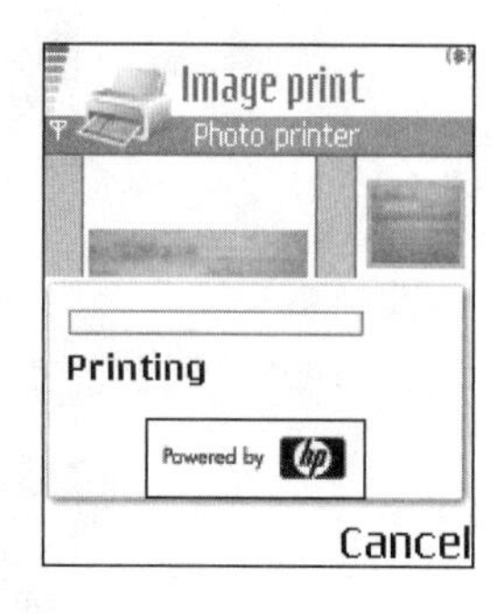

Windows Mobile Phone Launch Jetcet. Select File Printing in the main menu, and you will see a list of folders and files. If you have saved your photos, for example, in the My Pictures folder, open the folder and choose the photo you want to print. When the photo is displayed on the screen, open the menu and select Print to start printing (see Figure 16-5).

Figure 16-5

Print a photo from the Jetcet application.

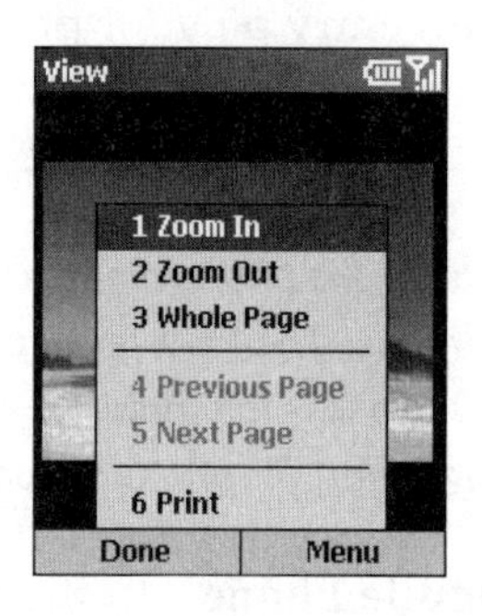

Step 2: Print from a Memory Card

A reliable technique for printing photos is to take the memory card from the camera phone and insert it into a card reader in the printer. This method is particularly handy if you have a printer with a small screen that lets you preview the images you are about to print.

All you need for this technique is a camera phone with a removable memory card and a printer with a memory card reader slot. Your memory card must be compatible with the printer, but the common SD and MMC cards are practically always compatible. The smaller variations of SD and MMC cards, such the Mini SD or RS-MMC, usually require an adapter for fitting them into a card reader.

All Phones Make sure your photos are saved on the memory card. If you need help, follow the instructions in Project 2.

Remove the memory card from your phone and insert it into the card reader in your printer. There may be several slots in the printer's front or side panel for different types of memory cards (see Figure 16-6). Find a slot where the card comfortably fits. Push it all the way to the end. If the printer doesn't react when you insert the card, check for a dedicated button on the control panel for memory cards or photos. You can also try to flip the memory card and reinsert it into the slot.

Figure 16-6
Insert your memory card into the printer.

There are two common techniques for managing printing from a memory card: you can view and select pictures on the display of the printer's control panel, or you can print out an index sheet of images and select photos from the sheet.

1. Print Using the Printer Control Panel

If your printer features a display that can show thumbnail photos, you can browse pictures stored on the memory card one by one and select the ones you want to print. The procedure for choosing pictures and initiating printing varies by printer model. Usually, you have to choose the paper size, photo size, and number of copies. Some printers let you modify lightness and darkness and color settings and may even let you crop the picture. Once you have marked all the photos you want, you can start the print job.

2. Print Using an Index Sheet

Printers that don't have a screen for displaying thumbnail images provide other techniques for selecting photos. Multifunction printers, which include scanner and copier features, can often create an index sheet of photos (see Figure 16-7). The index sheet is a collection of thumbnail pictures that the printer has discovered from the memory card.

When you insert your memory card into the printer, it reads all photos it can find from the card. Follow the printer's instructions for printing out the index sheet.

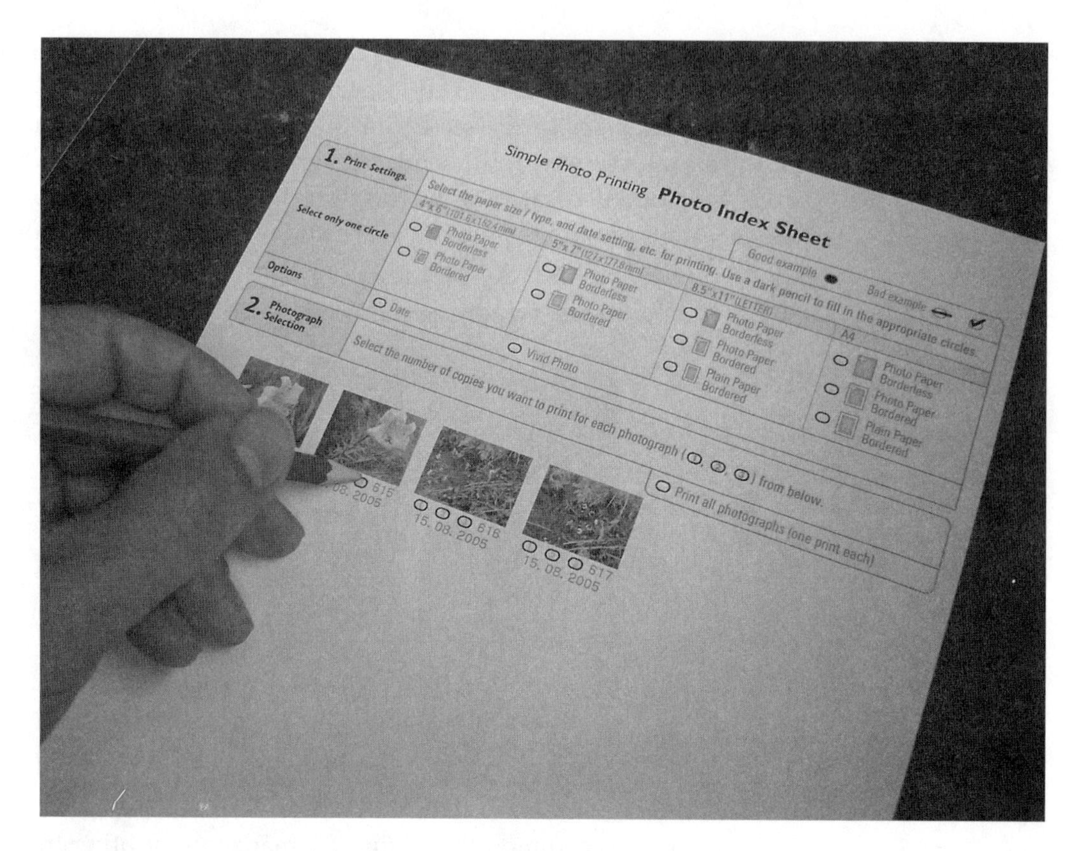

Figure 16-7

Select the images to print from an index sheet.

When you have managed to print the index sheet, mark the photos you want to print on the sheet. The index sheet may include the instructions, or find out from your printer's user guide the correct way of marking the photos. In addition to indicating which pictures are to be printed, you usually have to define the print size, number of copies, and paper type.

Once you have marked the photos, set the index sheet into the printer for scanning. The scanner reads the index sheet, matches your marks with the images on the memory card, and prints out the requested photos. Follow the instructions on the printer's control panel for confirming the print job.

Step 3: Print via a USB Cable

The key technology for connecting a digital camera or camera phone to a printer via cable is a standard called PictBridge. PictBridge has been specified by printer, digital camera, and phone manufacturers for allowing devices of any brand to print to any PictBridge-compliant printer.

Check your camera phone and printer user guides to determine if the products include the PictBridge technology. An up-to-date list of cameras, phones, and printers that have been certified as PictBridge-compatible can be found on the Internet at the

Camera & Imaging Products Association Web page www.cipa.jp/pictbridge/CertifiedModels/PictBridgeCertifiedModels_E.html.

1. Connect the Cable

To set up printing, hook up the USB cable from your camera phone to a compatible printer (see Figure 16-8). Switch on the printer.

Figure 16-8

Connect the USB cable from your camera phone to the printer.

2. Print

Symbian OS/S60 Phone In the main Menu of the phone, open the Gallery application (or Media, Gallery). Select Images and find the photo you want to print. Open the picture, and select Options | Print | Settings | Default Printer. Choose PictBridge Printer from the available options (see Figure 16-9). Select Back, and you should be back in the screen where you can initiate printing (see Figure 16-10). Then select Options | Print.

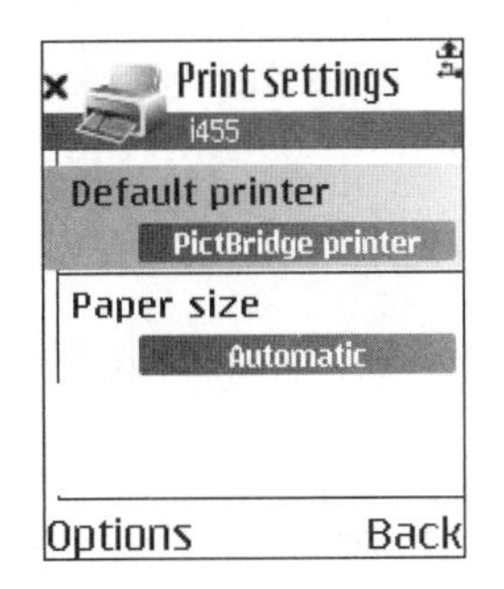

Figure 16-9

Define PictBridge printer as your printing device.

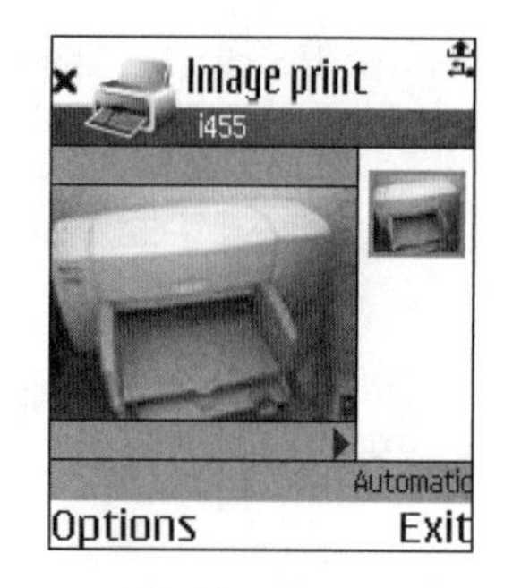

Figure 16-10
You can place the image on the paper before printing it.

In potential error situations, check both the printer and the phone for any error messages. Depending on your printer model, you may also have to use its control panel to start printing.

Make Backup Copies of Your Photos

Have you seen photos taken 20 or 30 years ago that have faded to an unrecognizable purple haze? It's such a shame that our memories disappear because of poor technology. The expected lifetime of a digital photo printed on an ink-jet or laser printer is even shorter than it was for those old film prints. You shouldn't only rely on paper copies for archiving your digital photographs.

The good news is that as long as pictures are stored in digital format, colors won't fade and the pictures won't be scratched. Digital photos are composed of bits and bytes that don't change (unless you edit the image on purpose). The only way to harm a digital photo is to lose it altogether. Hard disk crashes are known to happen, memory cards may refuse to cooperate, and CDs can be lost or broken. That's why you should save your pictures on more than one physical medium.

Here's the equipment you need for backing up your photos:

- Sufficient hard disk space on your computer for storing your photo library.
- CD or DVD copier. Recent computers come with a CD or DVD drive that can create or burn discs.
- Software for copying photos and other information to CD/DVD discs. This application is usually included with the CD/DVD disc drive.
- Blank CD or DVD discs. Two kinds of discs are available: R and RW. The difference is that you can copy information on an R disc only once, whereas an RW disc lets you add new information later on as well. R discs are more reliable, and they work on a wider range of disc players, so they are recommended for backup purposes.
- Waterproof pen for writing on the disc.

Take the following steps to back up your camera phone photos:

1. Copy your photos from your camera phone and from all your memory cards to your computer. See Projects 2 and 3 for detailed instructions.

2. Once the photos are saved on the hard disk, get blank CDs or DVDs. A CD can store up to 650MB of information. For example, a one-megapixel camera phone photo typically takes about 100–200KB of storage space. A good-quality two-megapixel camera phone photo takes about 300–450KB of memory space. You can back up about 1500 photos of 400KB size on a CD. A DVD can store up to 4.7GB (4700MB) of information.
3. Launch the CD/DVD copier software on your computer. Select data or photo disc when the application asks which type of disc you want to create. Also, you want to burn a disc that consists of files you pick up from the hard disk. Follow the instructions to burn a new disc.
4. Write down the backup date and other identifying information on the disc.
5. Always test the new CD before archiving it. If possible, try to read the CD on another computer and verify that the information you intended is saved on the disc.

Even CDs or DVDs won't last forever. Research has been done that suggests that a CD-R disc should last for 10–25 years when stored in a cool, dry place away from the sun. However, as long as the CD can be accessed by a computer, the information is intact and a new copy can be made.

Be Aware of Printing Costs

Color printing uses up plenty of color from cartridges, and the glossy photo paper is not that cheap either. You can print on normal document paper, but the result is not as nice as a printout on photo paper. Also, the colors tend to fade faster from document paper than from photo paper. If you calculate all the costs involved in a photo printed at home and compare it to a print by a photo lab, the results may surprise you.

For getting some idea of the costs involved in printing digital photos, let's see what a professional outfit is charging. At the time of writing this, competitive pricing with major photo print services on the Internet was about $0.19 US in the United States, or 0.15 euros in Europe for a 4×6-inch (10×15 cm) size photo.

Printing at home isn't necessarily any cheaper. Photos need plenty of color from print cartridges and you have to get glossy photo paper as well. It is possible to print two to four photos on one sheet of paper, if your printer and software allow it. Still, printer manufacturers have calculated that it costs around $0.28 US to print a 10×15 cm (4×6-inch) size photo on a home printer. It doesn't include the initial investment in the printer itself.

Ink-jet printers and multifunction (copier/scanner/printer) products are the best solutions for occasional home printing. These printers can be used for document printing as well. Photo printers can print higher-quality photos than ordinary or multifunction printers, but they are often dedicated to photo output only.

Project 17

Listen to Podcasts on Your Phone

What You'll Need:

- **A phone that can play MP3 audio**
- **A memory card on your phone and a memory card reader attached to your computer. Alternatively, you may use Bluetooth or a data cable for transferring podcasts to your phone**
- **A high-capacity memory card for storing podcasts. 512MB or 1GB memory cards have a good price/capacity value**
- **A podcast receiver application on your computer that can download new shows**
- **An advanced phone that can run a podcast receiver application allows you to download shows directly from the Internet**
- **Cost: $15–$50 U.S. for a memory card, depending on its capacity. $0–$20 U.S. for a podcast receiver application for a smartphone**

Podcasts are the talk shows of the digital age. Instead of watching a talk show on TV or listening to it on the radio, you can listen to the show on any computer or device that can play MP3 audio. The technical difference between the traditional shows and podcasts is that you can download a podcast to a portable device and listen to it wherever and whenever you want. Another difference compared to the traditional shows is the show itself. Anyone can produce a podcast—it only takes a microphone and a computer—and make it available on the Internet for anyone to download.

Perhaps it is the possibility of listening to both trusted professionals and passionate amateur broadcasters that makes podcasts so interesting. Podcasts may be entertaining, provide news, or teach a thing or two if they are educational shows. Professional producers, such as radio stations and magazines, deliver timely shows that have their facts straight. Amateur podcasters broadcast whatever they feel like saying. They can attract large audiences, as the huge popularity of personal blogs and shared photo albums, which existed before podcasts, proves.

First, we are going to find podcasts for you to sample. Then, you'll learn to subscribe (it's free) to the best ones and finally, you'll be able to download shows to your computer and to your phone.

Step 1: Discover Podcasts

Many major media companies and corporations produce their own podcast shows. Apart from the brand names, finding a podcast that you want to download and listen to is a game of luck. One strategy is to visit the Web pages you frequently read and find out if they provide podcasts. If you can find a show, simply download the related MP3 file and play it in a music player application on your computer or on your phone.

Another option, particularly handy if you are new to podcasts and want to get a clue what they are all about, is to visit a podcast directory. For example, Yahoo provides categorized listings of shows at podcasts.yahoo.com. You don't have to download the whole show before listening. The service can stream the podcast without you having to wait for the download to finish. Another podcast directory can be found at Podcast Alley (www.podcastalley.com) (see Figure 17-1).

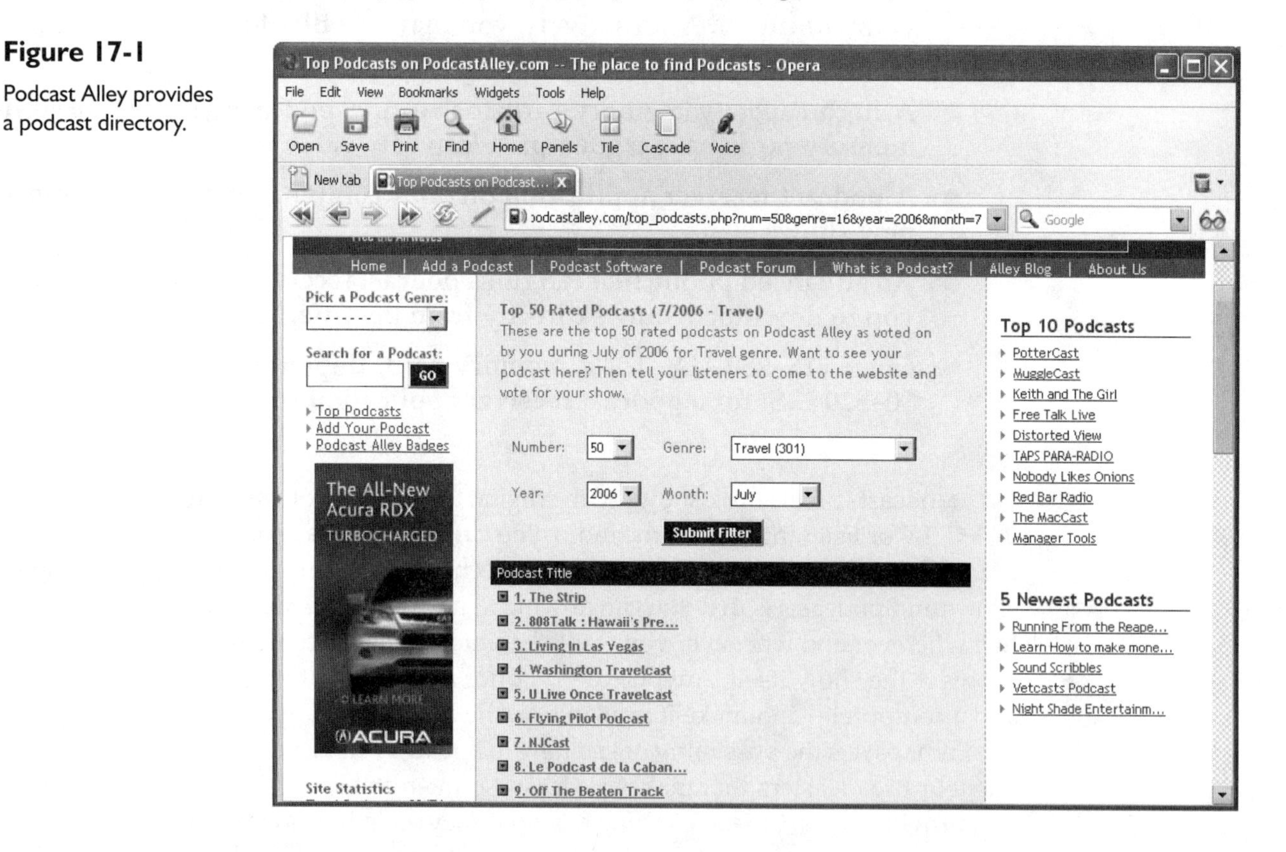

Figure 17-1
Podcast Alley provides a podcast directory.

Once you have discovered a few podcasts that you want to try out on your mobile device, you should get a podcast receiver application, also known as a podcatcher. The purpose of the application is to make the process of checking for new episodes and downloading them as smooth as possible.

You have two options for setting up a quick and easy podcast download process:

1. *Make your personal computer your podcast management center.* This setup allows you to listen to podcasts both on the computer and on any portable MP3 player that you may have, including your phone. First, you download new podcasts to your computer and then copy them to your mobile device.
2. *Turn your phone into a podcast receiver.* This setup allows you to download podcasts directly to the phone. You need a phone with fast data connectivity, such as EDGE, EV-DO, HSDPA, UMTS, or Wi-Fi access.

First, let's see how to create a podcast management center on the PC.

Step 2: Download Podcasts to Your Phone via Your PC

Users of Apple iTunes software already have an application for downloading podcasts; others have to install a new piece of software on their computers.

If you are an iTunes user, view the left pane of its main window for the label Podcasts. If you can't find the Podcasts section in your iTunes, update the software to a newer version. I'm using iTunes versions 6 and 7 in this project.

note *Keep in mind that although it is a comprehensive software package, iTunes can only copy podcasts and music to Apple iPod MP3 players and phones that come with iTunes software, like select Motorola phones. Still, it is possible to manually copy podcasts from your computer's iTunes library to any phone, as we will see later in this project.*

Another, more generic podcast download solution that works for all MP3-capable phones is a combination of two products: a podcast receiver application and Windows Media Player. I'll be using versions 10 and 11 of Windows Media Player in this project. The podcast receiver application is intended for downloading new shows, and Windows Media Player copies them to your phone or memory card.

Popular podcast receiver applications are, for example, Doppler (www.dopplerradio.net), FeedDemon (www.newsgator.com), and Juice Receiver (juicereceiver .sourceforge.net). Doppler and Juice Receiver are available as free downloads. I'll use Doppler release 3 beta in this project, but the principles for finding, subscribing, and downloading podcasts is similar in other products. If you are not using iTunes, install a podcast receiver application on your PC before proceeding in the project.

1. Subscribe to a Podcast

It is possible to manually check for new podcasts and download each new episode when they become available. If you intend to listen to more than one podcast, it is better to subscribe to interesting podcasts right away and unsubscribe if they prove to be something else you expected.

Doppler Doppler has a limited podcast directory. It is better to visit Podcast Alley or Yahoo Podcasts for searching shows. When you have discovered a podcast you like, click the Subscribe link at Podcast Alley. If you are in Yahoo Podcasts, find the RSS link of the podcast. You want to find the Internet address (URL) of the podcast. Select the address and choose Edit and Copy from your browser menu.

Switch to Doppler. Click the Add button at the bottom of the window. You should see the same address that you just copied from the podcast directory in a small window (see Figure 17-2). Confirm your subscription by clicking the Finish button.

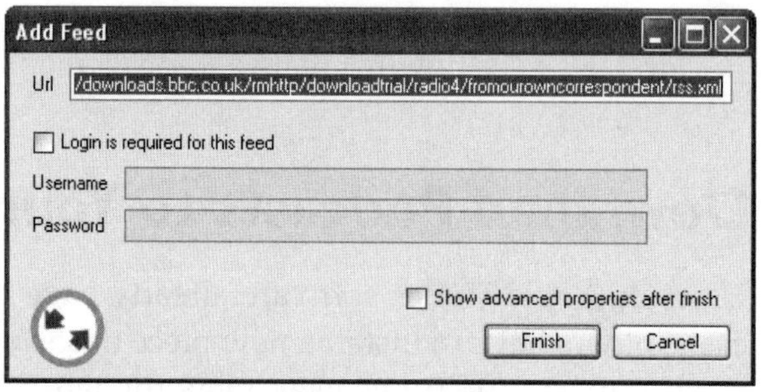

Figure 17-2
Subscribe to a podcast in Doppler.

Repeat the procedure for each podcast you want to subscribe to.

iTunes Launch iTunes. Click Podcasts in the left pane. Then select the Podcast Directory link at the bottom of the window. Scroll down the main window, and you can view podcast categories. Browse or search podcasts (the search box is at the top). You can listen to a podcast before downloading it by selecting a show, and clicking the large Play button at the top-left corner of the iTunes window. When you have discovered a podcast you want to subscribe to, click the Subscribe button and confirm your subscription (see Figure 17-3).

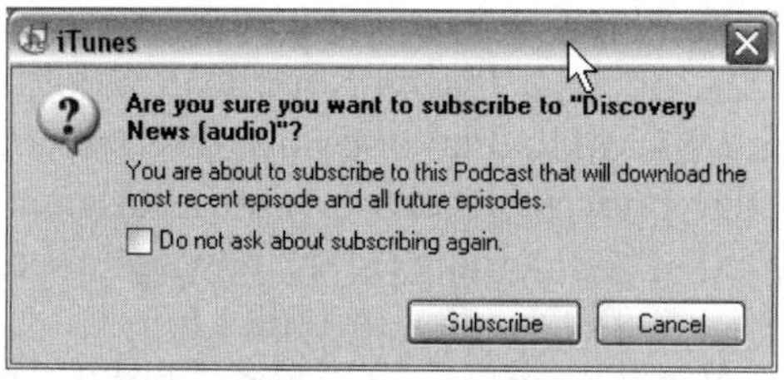

Figure 17-3
Subscribe to a podcast in iTunes.

Repeat this for each podcast you want to receive on your PC.

2. Define the Download Details

Once you have subscribed to a number of podcasts, and before doing anything else, you should configure a few things in your podcast receiver application to make your downloads easier.

Doppler Define a download folder in Doppler that the Windows Media Player can also find. If your music library folder in Windows Media Player is, for example, C:/Music, create folder C:/Music/podcasts for downloaded podcasts (see Figure 17-4).

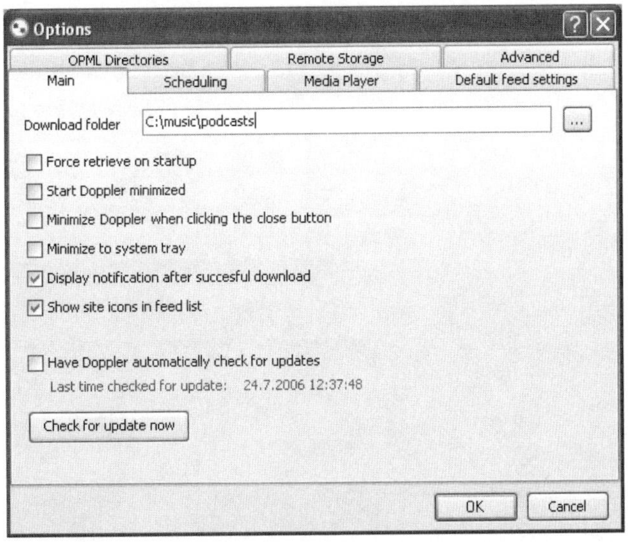

Figure 17-4

Create a folder for podcasts under the Windows Media Player music library folder.

Open Tools in the Doppler menu and select Options. Click the button with three dots beside the download folder, and select the folder where downloads will be saved.

Before you start downloading (or retrieving, as Doppler calls it), you have to catch up with earlier shows. Maybe the podcast you have subscribed to has already published hundreds of episodes. Since you haven't downloaded anything yet, they are all new for you and Doppler believes it has to download all shows. Define how many of the old shows you want to download.

Open the Edit menu. Select Catch-up All Selected Feeds. If you want to download only the latest episode from all the podcasts you have subscribed to, enter **1** into the Leave Last Podcasts field (see Figure 17-5).

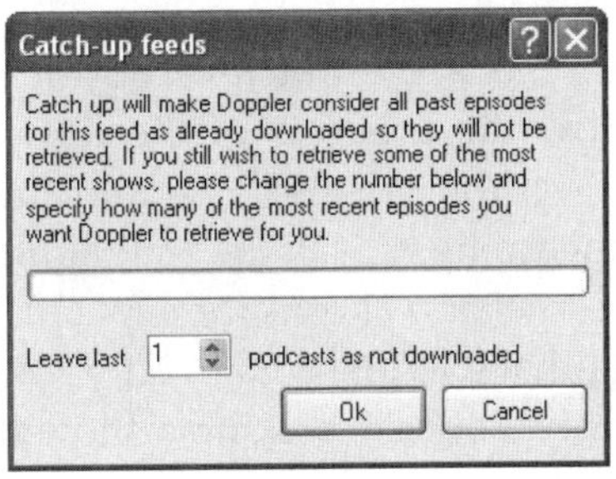

Figure 17-5

Define how many old episodes you want to download.

After the first download, you don't have to worry about catch-up anymore, because Doppler keeps track of new podcasts and which ones you have downloaded.

iTunes List your podcast subscriptions by clicking Podcasts in the left pane. Click the Settings button at the bottom of the window. Here you can change the settings for your podcasts (see Figure 17-6). For instance, Download the Most Recent One ensures that iTunes only downloads one show per podcast. You may consider changing the value of the Keep-field to All Unplayed Episodes in order to save disk space. Otherwise, iTunes will keep all old shows on your hard disk.

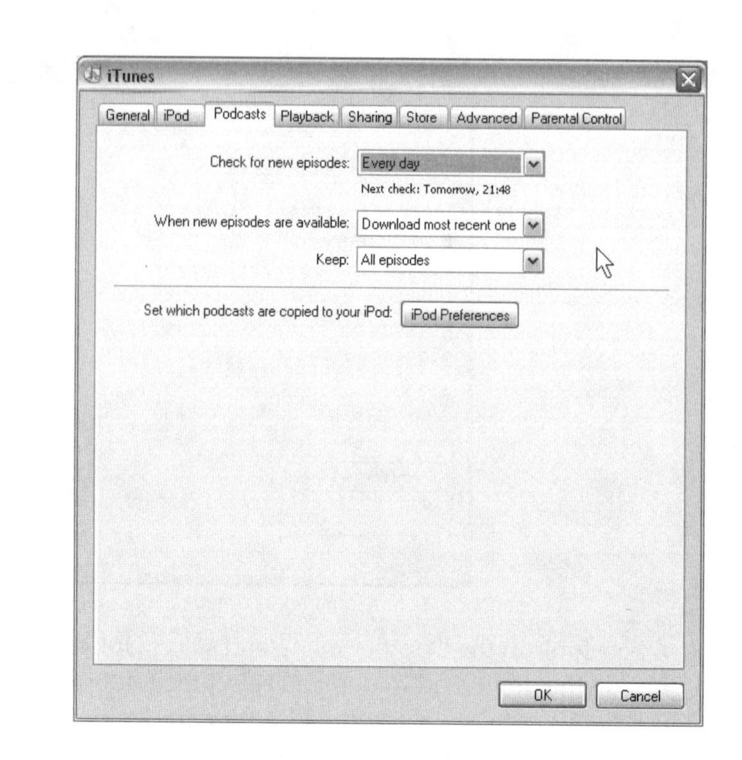

Figure 17-6

Podcast settings in iTunes.

3. Download Podcasts to Your Computer

Podcast receiver applications can automatically retrieve new episodes at defined intervals or you can initiate the download process, for instance, when there's no other load on your computer.

Doppler Click the Retrieve button at the bottom of the Doppler window. You may follow the progress of the downloads at the lower half of the window (see Figure 17-7). You can also schedule automatic downloads. Open the Tools menu and select Scheduling. Check the Scheduling box. Define the times when Doppler logs on to the Internet and checks for new episodes.

If you have loudspeakers on your computer, you can listen to podcasts in Doppler. Select a podcast, pick up a show in the top-right pane, and click the play button in the pane on the right.

iTunes iTunes begins automatically downloading the latest episode when you subscribe to a podcast. The next time you are in iTunes, you can click the Refresh button at the lower right hand corner (Update button at the top-right corner in release 6) to initiate downloads. You can also define an automatic download schedule in the iTunes Settings.

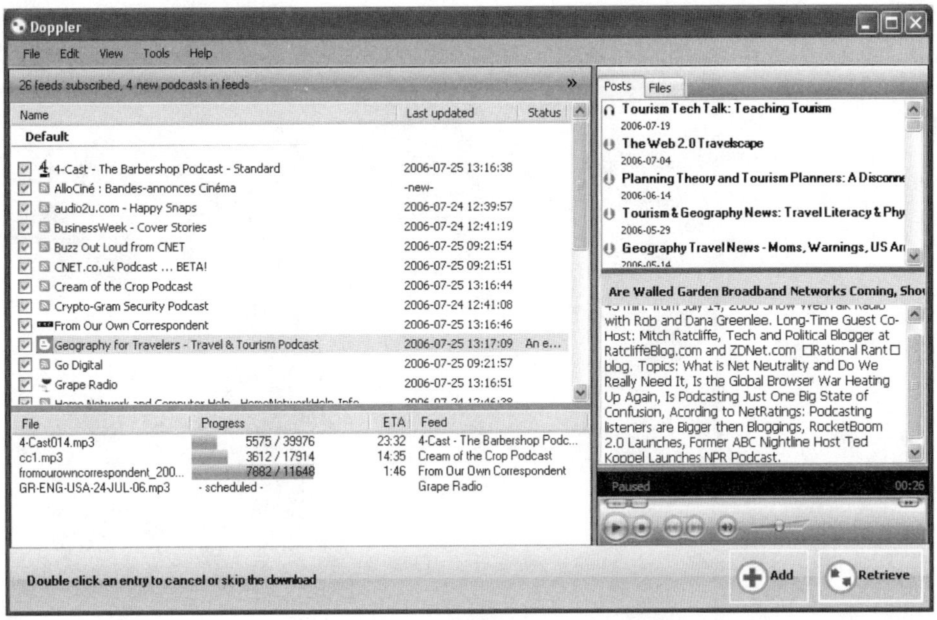

Figure 17-7

Downloading podcasts in Doppler.

4. Set up a Sync between the Media Library and Your Phone

Now that you have saved the latest podcasts on your computer's hard disk, set the media player application ready (Windows Media Player or iTunes) for copying the shows to your mobile device.

Windows Media Player Launch Windows Media Player. The application identifies the new downloads as podcasts, assuming that you created the Doppler podcast folder under the Windows Media Player's music library folder.

Click the Library tab in the menu bar at the top. In the left pane, click Library and Genre. Scroll down until you find Podcasts. Some podcasts may have been categorized as news, sports, or technology (it is up to the broadcasters how they label their shows).

You have to let Windows Media Player know which podcasts you want copy to your phone or memory card. This is done by creating a playlist. Drag and drop your podcasts into the pane on the right (see Figure 17-8). Click the Save Playlist button at the bottom of the window, and enter a name for the new playlist.

If you can't find the Save Playlist button, open the program menu. Then, select File | Save Now Playing List As. Type a name for the new playlist. It will become your playlist for syncing podcasts to the phone.

iTunes You can find the sync settings for podcasts by opening Edit in the main menu. Select Preferences | iPod | Podcasts (see Figure 17-9).

note *iTunes can only sync with Apple iPod devices and phones that come with iTunes software. Nonetheless, if you want to use iTunes for discovering and downloading podcasts, but you don't have a compatible phone, you can manually copy the MP3 files using Windows Explorer.*

Figure 17-8

Create a Playlist for the podcasts you want to transfer to your phone.

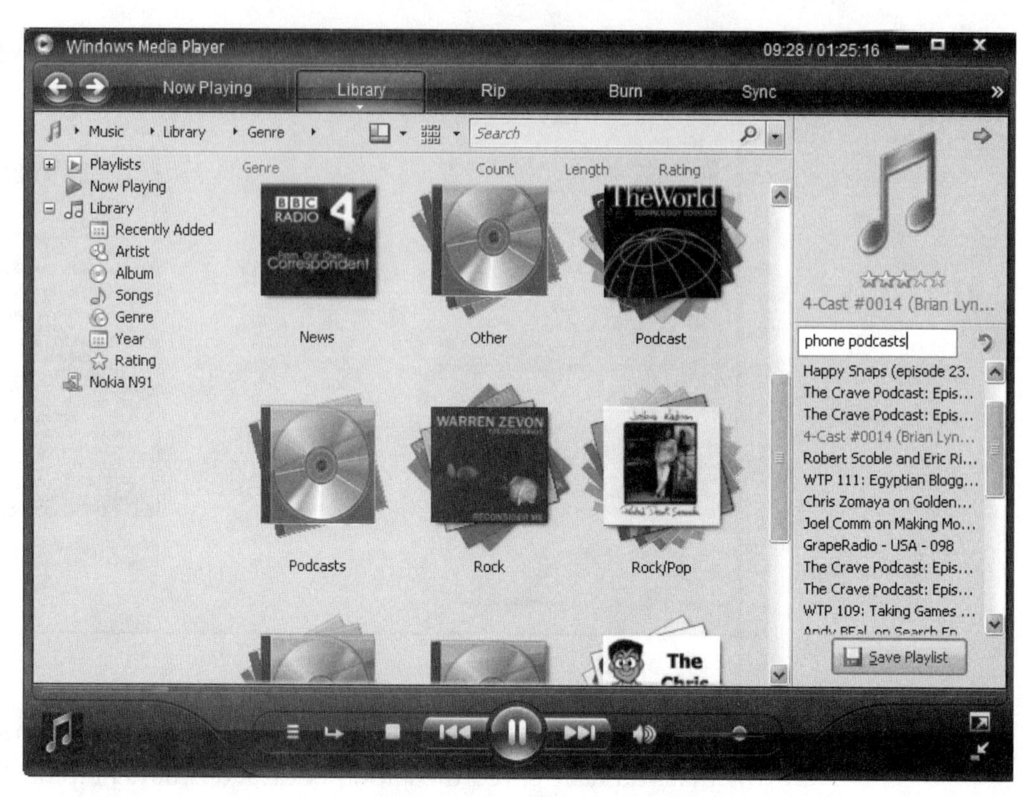

Figure 17-9

Set up sync for podcasts in iTunes.

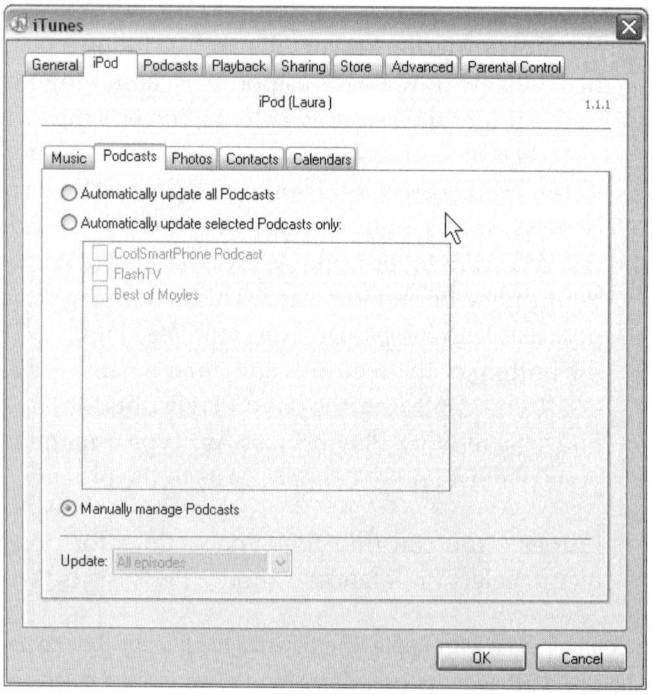

5. Copy Podcasts to Your Phone

If you want to use the sync feature in Windows Media Player or in iTunes, you must have a data cable connection between your phone and PC. Windows Media Player can also sync with a memory card.

Windows Media Player Insert the memory card from your phone into a memory card reader attached to your computer, or hook up the data cable from your phone to the computer. Pop-up windows may appear on the screen asking what you want to do with the device. If this is the first time you are connecting the device to this computer, Windows may ask your permission to configure itself for the device. Let it do so.

Windows Media Player 11 identifies the connected device and calculates the available memory space on the unit. You can view the memory status in the pane on the top-right corner (see Figure 17-10).

Figure 17-10

Monitor the memory usage.

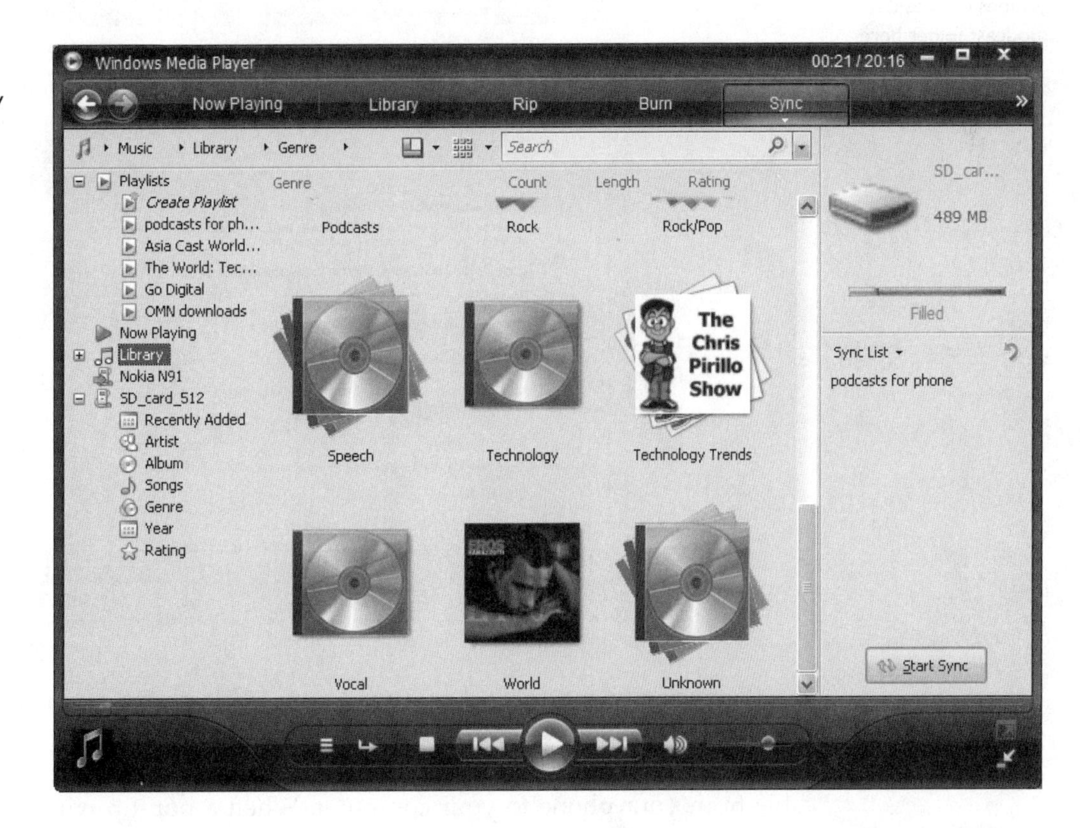

In the leftmost pane, open Playlists. Find the playlist you created earlier. Grab the playlist with your mouse, and drag and drop it into the right pane under the title Sync List. If the bar displaying the remaining memory space turns red, all items included in the playlist won't fit into the device. Still, you can click the Start Sync button at the bottom of the pane. Windows Media Player will copy as many podcasts as it can fit into the available memory space.

Windows Media Player 10 requires you to select Edit Playlist and click each podcast you want to sync. Click OK once all podcasts are listed in the right pane. Select Start Sync to copy podcasts.

When the sync process has finished, unplug the data cable. If you used a memory card, remove it from the reader and insert it into the phone.

iTunes If you have an iTunes-compatible phone, plug the cable from the device into the computer. iTunes will recognize the device and start syncing automatically if you have defined it so in Settings.

Find out where iTunes has saved the podcasts. In the main menu, select Edit | Preferences | Advanced | General. Then iTunes Music Folder Location displays the folder where the downloaded podcasts are stored (see Figure 17-11).

Figure 17-11

iTunes creates its podcast folder here.

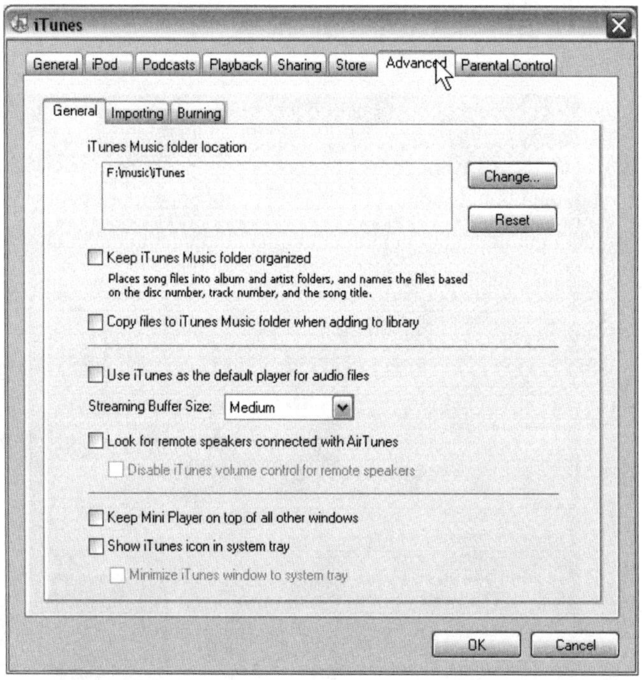

If you don't have an iPod or iTunes-compatible phone, you need additional help from Windows Explorer.

Insert the memory card from your phone into the card reader or plug in the data cable from your phone to your computer. When a pop-up window appears on the screen, select Windows Explorer. Using Windows Explorer, locate the podcast folder on the hard disk. Drag and drop the podcast folder to your phone or memory card (see Figure 17-12).

note *Windows Explorer copies as many podcasts as it can fit into the storage space. If the space runs out, it stops. Then, you have to manually delete other files from the memory card. Be very careful if you do this. Don't touch any files or folders that you don't recognize as your own.*

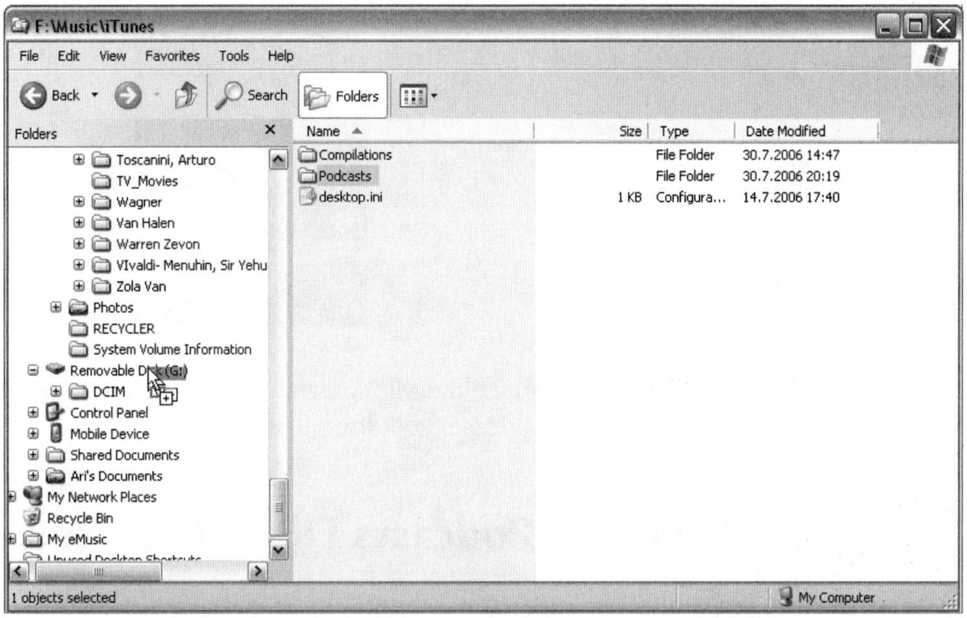

Figure 17-12

Copy podcasts from iTunes to your phone.

6. Listen

Because podcasts usually are MP3 files and digital music is also often stored in MP3 format, the music player application on the phone may believe that podcasts are music as well. That's fine. It only means that you may have to search for podcasts among song titles.

Symbian OS/S60 Phone Launch the Music Player application on your phone. You can find it in the main menu or in the Media folder. To make sure that the new podcast titles are listed in the Music Player, make the application scan the storage space for new items. Select Options | Update Collection (or Refresh Music Library) | All Songs and find your podcast (see Figure 17-13). Press the selection key to listen.

Figure 17-13

Play a podcast.

Windows Mobile Phone In the Home screen, press START. Locate the Windows Media application and open it. If your phone displays all tracks right away, highlight a podcast and press PLAY. If your phone displays a list of media categories, choose Music | Genre | Podcast (see Figure 17-14). This way, you don't have to view any other items than podcasts. Highlight a podcast and press PLAY.

Figure 17-14

Play a podcast on a smartphone.

Other Phones Launch the MP3 player application on your phone and list all tracks. Locate one of the podcasts you downloaded and play it.

Step 3: Download Podcasts Directly to Your Phone

One of the reasons why a modern phone with a high-speed 3G or Wi-Fi network connection is so convenient to use is that when you want something from the Internet, you can get it right there and right then when you want it. This applies to podcasts as well. You don't need a computer for downloading and listening to podcasts at all, because you can get the shows directly to your phone from the Internet.

note *A typical podcast lasts about 10–50 minutes. To receive a podcast over the air to your phone, you have to download an MP3 file of about 5–30 MB in size. A fast network connection, such as EDGE, EV-DO, HSPDA,UMTS, or Wi-Fi is required for achieving this. For example, it only takes a minute or two to download a short podcast over a Wi-Fi connection, whereas downloading a 30-minute show over an EDGE connection will take 15 minutes or longer.*

note *In addition to the fast wireless network connection, you need a generous data communication plan that doesn't overload your budget. You may download tens, or hundreds, of megabytes of information each month if you really get into podcasts. A flat-rate data subscription plan can be a worthy choice. Wi-Fi is different because you may access the Internet through an access point at your home, office, or coffee shop for free.*

1. Install a Podcast Receiver Application on Your Phone

Only a few phones come with a built-in application that lets you download podcasts. Fortunately, a variety of podcast applications are available from software vendors for many phone models.

Symbian OS/S60 Phone I'm going to use a product called Mobilcast from Melodeo on a Symbian OS/S60 device in this project. Mobilcast is a Java application that runs on many types of phones regardless of their operating system software. Mobilcast is a free application. Other podcast applications for S60 mobile devices are, for example, Mobiedge Radio (download available at www.handango.com), Nokia Podcasting (www.nokia.com/podcasting), and Voiceindigo (www.voiceindigo.com). Some are free, but some products require you to purchase a license after the trial period has expired.

Go to the Melodeo's Web page at www.melodeo.com. Register with the Mobilcast application. During the registration process you will receive a text message with a link to the download, or you will see the download link that you can enter into your phone's Internet browser. I downloaded the application from zeus.melodeo.net/mc, but you should verify the address during your registration.

After the download, installation will begin automatically and all you have to do is confirm that you really want to install it. When the installation routine has finished, you can find Mobilcast (see Figure 17-15) in the My Apps, Install, or Fun folder.

Figure 17-15

Mobilcast application provides a podcast directory and also the actual shows.

Windows Mobile Phone I'm going to use an application called SmartFeed as the podcast receiver on a Windows Mobile smartphone (see Figure 17-16). The product is available at Newsgator SmartFeed (www.smartfeed.org). Other podcast products for Windows Mobile smartphones are, for example, SmartRSS for Windows Mobile Smartphone (www.beetzstream.com) and Skookum Subscriber (www.skookummobile.com).

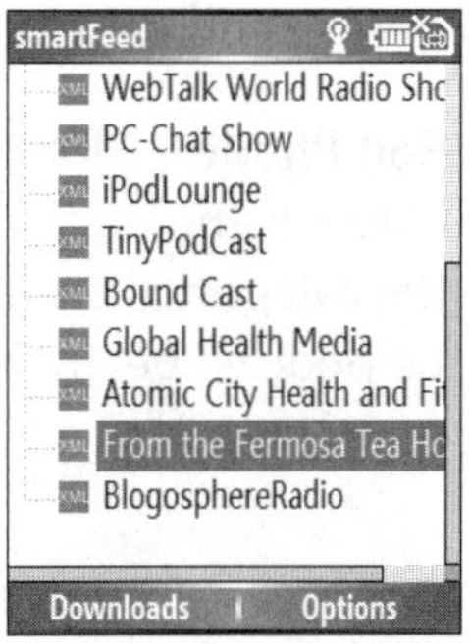

Figure 17-16

SmartFeed podcast receiver application lets you subscribe and download podcasts directly to your Windows Mobile phone.

On your phone, point the Internet Explorer browser to www.smartfeed.org. Select the version of the software that is compatible with your phone. When the download is finished, the installation will begin automatically.

If you have problems downloading the application directly to your phone, you can always log on to the same Web page using your computer and download the file onto your hard disk. Then, connect the data cable from your phone to the computer and install the application.

Once the application is installed, go to the Home screen, press START and find SmartFeed from the menu. Launch SmartFeed. You should check one setting before doing anything else. Open the Options menu and select Settings. Make sure that the Podcast Directory points to a folder on your memory card (see Figure 17-17).

Figure 17-17

Define a folder on the memory card for downloads.

Other Phones Go to the Melodeo's Web page at www.melodeo.com. Register with the Mobilcast application. Follow the instructions outlined for the Mobilcast application. Mobilcast is compatible with a variety of phones. Even if during installation the Web page informs you that your phone may not be compatible, it can be worth trying if you know that your phone can run Java applications.

During the registration process you will receive a text message with a link to the download, or you will see the download link that you can enter into your phone's Internet browser. I downloaded the application from zeus.melodeo.net/mc, but you should verify the address during your registration.

2. Subscribe to Podcasts

When you subscribe to podcasts, the podcast receiver application can automatically check for new shows and download them for you.

Symbian OS/S60 Phone Mobilcast doesn't require you to subscribe to podcasts at all. When you launch the application, it displays a directory of podcasts sorted by topic. Browse the categories until you find a podcast you want to listen to. When you select Listen, the podcast begins to play almost instantly (see Figure 17-18).

Figure 17-18

Listen to a podcast in Mobilcast without downloading the file.

note *Mobilcast uses streaming technology to deliver podcasts without downloading. The benefit is that you don't have to wait for the download to finish, but streaming requires a network connection for the whole duration of the podcast. Also, the audio quality of a streaming podcast tends not to be as good as the audio quality of a downloaded podcast.*

Windows Mobile Phone In the main screen of the SmartFeed application, select Options | Add Feed. Click the Podcasts label in the tree structure. You can browse the podcast directory, for example, by category (see Figure 17-19). When you discover a podcast you'd like to subscribe to, press the selection key. Repeat the process for each podcast you want to download.

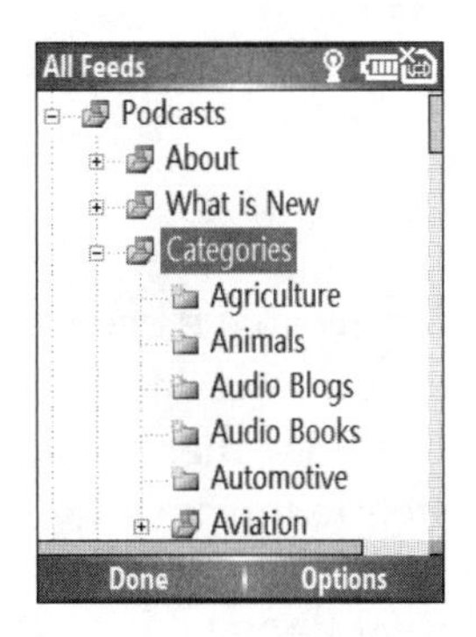

Figure 17-19
Browse the podcast directory in SmartFeed.

Other Phones Mobilcast users don't necessarily have to subscribe to podcasts at all, because you can listen to shows without downloading (although it is possible to download as well). When you discover a show you want to listen to, select Listen at the bottom of the screen.

3. Download Podcast Shows to Your Phone

Let the podcast receiver application manage your downloads, but don't let it automatically initiate the retrieval of new shows. This is because you may want to download only, for example, when you are at home and have access to your Wi-Fi network.

Symbian OS/S60 Phone It is possible to download podcasts in Mobilcast as well. You have to do it manually, one podcast at a time. At the bottom of each podcast episode you should see Save For Later. Mobilcast will establish a separate connection using the phone's browser software to download the podcast. Once it is downloaded, remember to save the podcast on your phone.

Windows Mobile Phone Once you have subscribed to a few podcasts in SmartFeed, you can download some episodes. Go to the main screen and select Downloads. At the top of the screen, you will see the podcast title. Push the navigation key to the right or to the left to switch to another podcast. When you want to download, scroll down the list and highlight an episode. Open the menu and select Start (see Figure 17-20).

Figure 17-20
Start the download in SmartFeed.

Other Phones You may download podcasts with Mobilcast by selecting Save For Later.

4. Listen

Once the new shows have been downloaded, you may listen to them whenever and wherever you want. After listening, immediately remove the episodes that you don't need to save for some other purposes. Podcasts take a considerable amount of memory space, and before the next download you must have space for new shows.

Symbian OS/S60 Phone Launch the Music Player application. It is located in the main menu or Media folder. Open the Options menu in the Music Player and select Update Collection (or Refresh Music Library) | All Songs. Find the podcast from the list and press PLAY.

Windows Mobile Phone In the main screen of the SmartFeed application, open the Options menu and select Manage Podcasts. You will see a list of downloaded podcasts. Highlight the one you want to play, open Options, and select Play.

Other Phones Launch the MP3 music player application. Find the downloaded podcast and play it.

Part III

Advanced

Project 18

Watch TV on Your Phone

What You'll Need:

- **TV tuner for your computer**
- **Fairly recent home computer that has the processing power for live TV and for streaming video and audio to the Internet**
- **Orb media-sharing software for your home PC**
- **Internet browser software on your phone**
- **Media player software on your phone, such as RealPlayer or Windows Media Player, that can play media streamed over the Internet**
- **Cost: $50–$150 U.S.**

While TV screens at home are growing larger, small mobile devices are making it possible to carry TV in your pocket, with exactly the same channels as you can view on your home TV. As futuristic as it may sound, you can build your own media and TV sharing center for a fairly small budget. The system is built on your home PC and broadband connection. Once you get your media-sharing center up and running, you can remotely access its resources from any device, be it phone, PDA, game console, computer, or laptop, as long as the device can access the Internet using a Web browser.

In this project, we will build Orb media-sharing system that lets you watch your home TV on your phone (see Figure 18-1). The same system also gives you remote access to your photo albums, MP3 music, and home videos. We, however, will focus on building a remote TV system that you can watch at the neighborhood shopping mall or at a bistro along the Loire valley. Not to mention watching TV shows you have recorded earlier by pushing a button on your phone.

note *In addition to Orb, which is a free product, there are other products that let you view your home TV on a phone. For example, Slingbox is a piece of hardware that you hook up between your TV and the Internet broadband connection. Using a compatible Windows Mobile device, you can then connect to the Slingbox over the Internet to watch your TV channels. Sony LocationFree makes it possible to watch home TV over the Internet on a Sony PSP.*

Figure 18-1

Watching home TV programming away from home can be an unforgettable experience.

Step 1: Choose a TV Tuner for Your PC

When shopping for a TV tuner that you intend to connect to your computer, it seems that there is an overwhelming selection of products. In reality, only some of the tuners are compatible with your TV system. If you don't already have a tuner or don't know which product to purchase, you should find answers for the following questions:

1. *Which TV technology do you have at home?* Is there an antenna on the roof, do you receive a satellite signal, or are you hooked up to cable TV? Is the TV broadcast signal analog or digital?
2. *Are you going to install the TV tuner in a laptop, or in a desktop computer?* If you want to install the tuner in a laptop, an external box with a USB connection is your choice. Desktop computer owners may consider installing a tuner card that goes inside the PC as well.
3. *Which TV tuner is compatible with your TV signal, PC and Orb?* Orb has tested a number of TV tuners that it supports. If you want to get a product supported by Orb, check the company's Web pages (www.orb.com/support) for the latest product list. Nonetheless, I installed a DVB-T tuner (Anysee E-30) that's not supported by Orb, because I wanted to get digital TV. This product works fine with Orb, along with other products Orb users have tested. View the Orb community forums at forums.orb.com/community for more information on products.

Step 2: Install the TV Tuner

Before installing the TV tuner, there are a few things to consider.

- Be prepared to extend the TV signal cable from your TV to your computer.
- In some cases, a simple cable splitter is handy. It allows you to split the TV signal cable both to your TV and your computer.
- If you intend to use a remote control for channel surfing, you need to think about the placement of the tuner box as well (if you have an external unit).

Follow the instructions you received with the TV tuner for installing the included software. Software for these products tends to be frequently updated. Check the manufacturer's Web site for the latest updates before installing the software from the CD.

Install the TV receiver hardware according to the user guide. If you have a card, you have to open the PC case and install the card into a free expansion slot. External units are easy to install—you simply connect the box to a USB port (see Figure 18-2) on your PC. It must be a high-speed USB 2 port because live video requires fast data transmission capability.

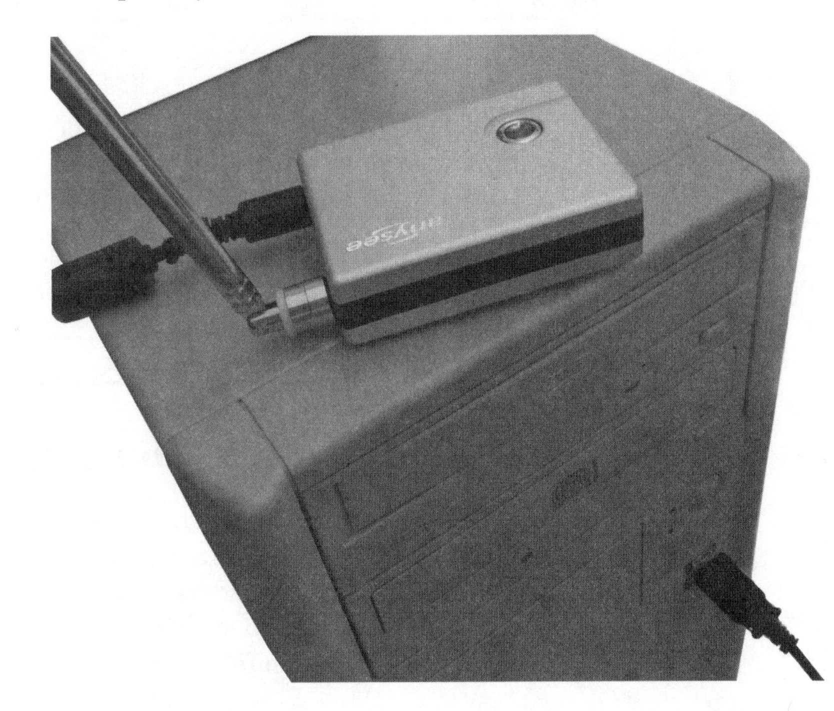

Figure 18-2
A digital TV tuner connected to a USB 2 port.

A computer installation is not ready before it is tested and proven to work. Now is the time to test your TV tuner. Follow the user guide to tune the TV channels with the included software. Make sure you can view the TV locally on your computer before moving on to the next steps.

Step 3: Download and Install Orb

Point your Web browser to the Internet page www.orb.com. Check that your computer is compatible with the software and has the resources for running Orb.

Click the Download link and choose your version of the software. The difference between the country versions is the online TV guide and language. If your language is not available and you are outside the United States, you might choose Canada or Great Britain; otherwise the U.S. TV guide may complicate things later. Follow the instructions for installing Orb.

When you have finished the Orb installation, it will start by itself. You have to create an account with a user name and password so that only you (and those family members you let know of the password) can have remote access to the computer (see Figure 18-3).

Figure 18-3

Create an account for remote access.

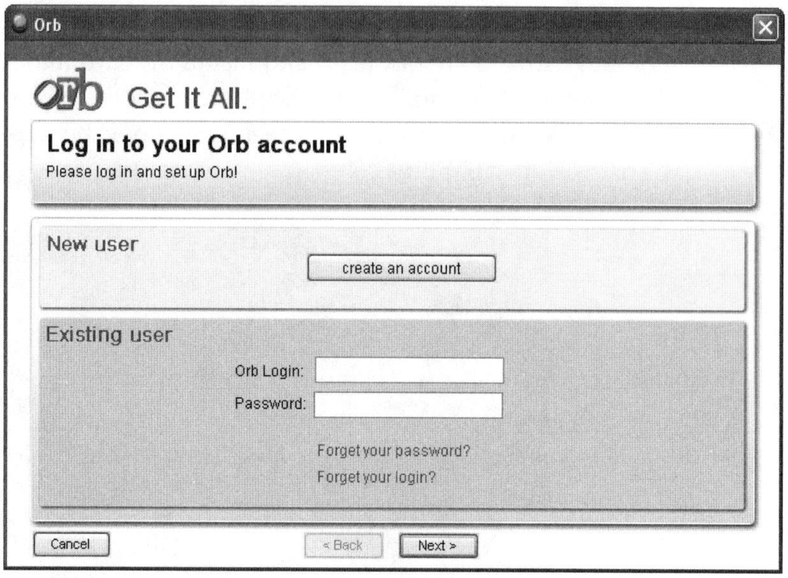

Step 4: Configure Orb for Media Sharing

Orb automatically searches your computer for folders that are likely to contain images, videos and music. It can also recognize installed Web cams and TV tuners. Orb won't necessarily find all your media folders. You can specify where the system can locate more folders. Right-click the Orb icon (green ball) on your Start menu toolbar (see Figure 18-4). Select Configuration from the pop-up menu.

Figure 18-4

Access the Orb configuration.

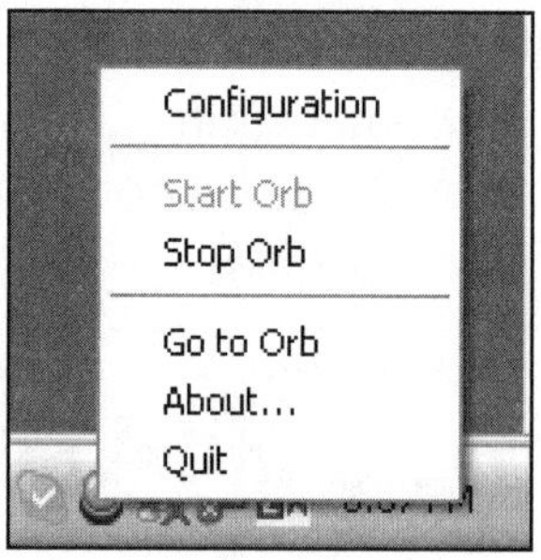

Select your login name. You can create more user names later with restricted access, for example, for your friends. You should now see the configuration window (see Figure 18-5) where you can tune the TV and specify your media folders.

For sharing your video, photo, and music folders, click the Media tab. Select the Add button to add a folder that is, for example, located on another hard disk. Highlight a folder and select Remove for media files you don't want to share (the files and folders won't be deleted from your hard disk).

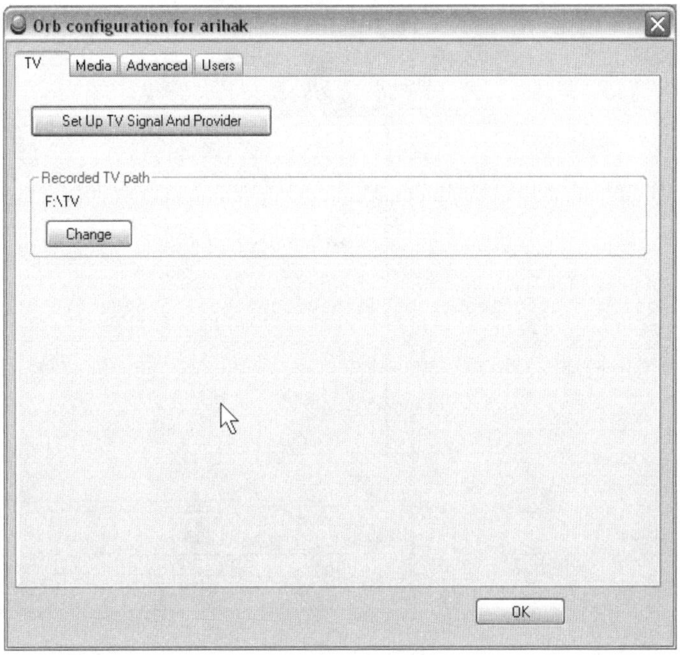

Figure 18-5
Define your TV channels and media folders.

Step 5: Tune the TV Channels for Orb

Although you already have tuned the TV, you must tune it once more for Orb. Click the TV tab in the Orb Configuration window (see Figure 18-5).

Click the Set Up TV Signal and Provider button. Choose the type of the connection that feeds the TV signal to your computer. For instance, I selected "I don't have a set-top box," because my digital TV tuner gets its signal directly from the outdoors antenna.

After selecting your country, you have to specify whether you use cable TV or a signal received by an antenna. Orb 2 may ask questions about your location for connecting you to a service that provides electronic program guides (EPG). Then, you should see the channel tuning window (see Figure 18-6). When you select the Start button, Orb will start scanning for signals. Choose Save Channel for channels you want to save, and select Find Next Channel to scan the next one.

Orb lets you know when it has scanned the entire signal range. At this point, you can close the tuner window.

Step 6: Test Your Remote TV

Let's make sure that your remote TV access works. The best way to test it is to watch TV on another computer. This can be a computer right beside your Orb server or at your friend's house. If you don't have access to another computer, use the same one where Orb is installed.

Launch the Internet browser on a remote computer and log on to your Orb account. Orb release 1 users have to go to my.orb.com in order to login to their account. Orb 2 users go to mycast.orb.com.

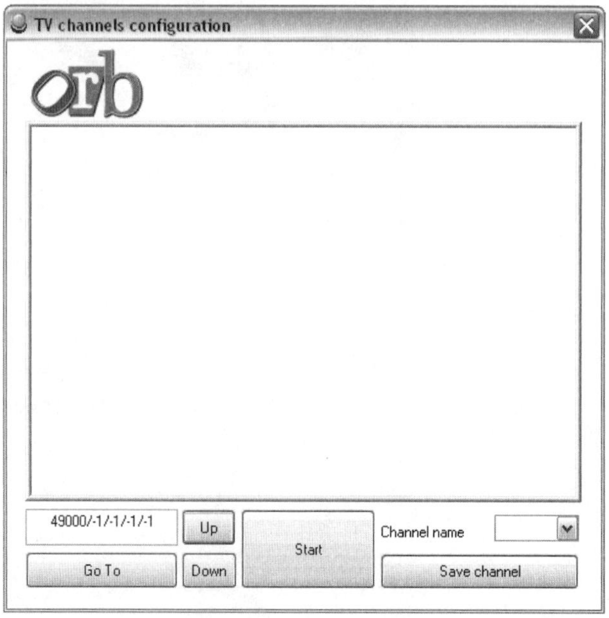

Figure 18-6

Scan and save the discovered TV channels.

If your home PC is running Orb 1, click the TV tab. Select Local TV Listings, and you should see the same channels you saved during channel selection. Select a channel, and you'll see a few control buttons (see Figure 18-7). Click the star button to save the channel to your Orb favorites, the REC button to start recording, or the Play button to watch the channel.

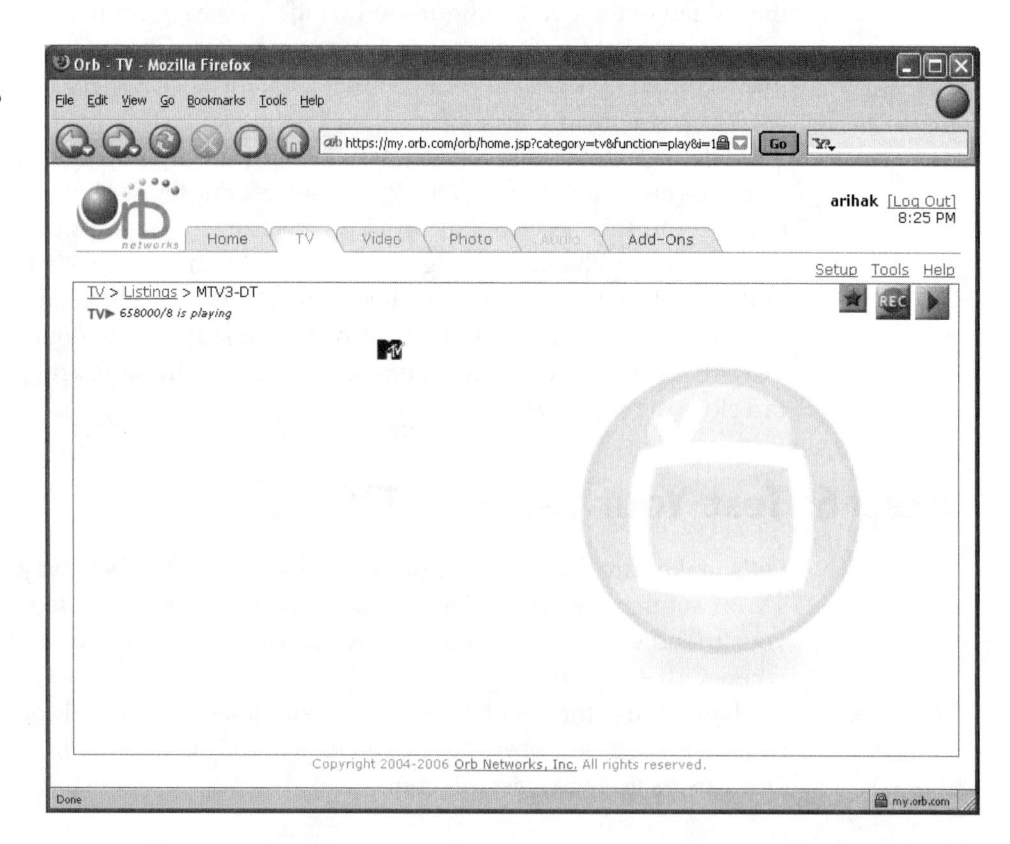

Figure 18-7

Click the Play button to watch a TV channel in Orb release 1.

If you are running Orb 2, click the TV – Program Guide pane. You can select a TV channel from the program guide.

When you click the Play button, Windows Media Player (or another media streaming application that may have been configured on your computer) will launch. The media player should play the video stream from the Orb server. Be patient, it will take a moment before you can see the picture.

tip *If you see a choppy TV picture that tends to pause and restart, one of the reasons could be a lack of processing power on your Orb server. Video decoding, encoding, and streaming require a fair amount of resources from the computer. Shut down all applications that you don't use on your Orb server. For example, you may close Skype (or any other Internet telephone application) and music and video player applications, such as iTunes and Windows Media Player. Many applications are hiding as small icons at the end of the Start menu toolbar. One by one, right-click each one. If you don't need the application just now, close it. Common examples of such applications are Quicktime, ActiveSync, Instant Messaging, and CD writers. Often, they are just lying around, waiting for something to happen so that they can quickly react. In this case, they are unnecessarily occupying memory space and take up processor time when you need it for something else.*

Step 7: Set Up Your Phone

If you have gotten this far in this project, you have successfully set up a media-sharing system that you can access practically anywhere, on any device. Next, let's set up your phone for remote access to Orb resources. Some mobile devices don't require any changes to their current setup for Orb access, but some phones require changes to the media streaming settings.

Symbian OS/S60 Phone Access to video and audio requires the cooperation of two applications: the Web browser and the media player on the phone. You have to make sure that these two applications play together, using the same Internet access point.

Go to the main menu, and launch the Web (or Services) browser application. Open Options and select Settings. If the access point is the one that you successfully have used for Internet access, remember its name (see Figure 18-8). The actual access point name can be anything—usually your service provider has labeled them.

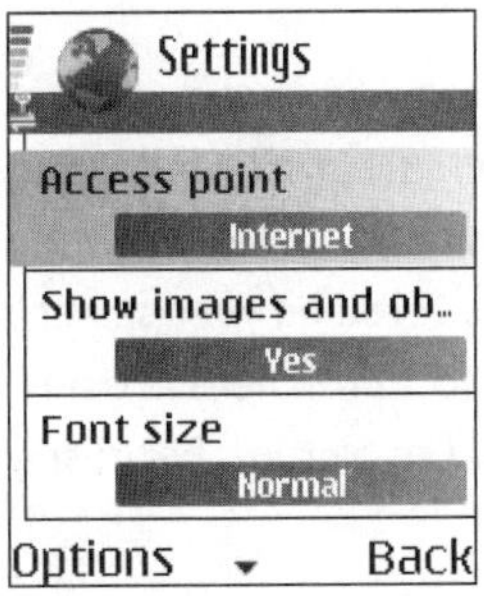

Figure 18-8

The Internet access point in the Web browser.

Return to the main menu and open the Media (Imaging) folder. Launch RealPlayer, and select Options | Settings | Connection and Network. Make sure the default access point is the same as that in the browser (see Figure 18-9).

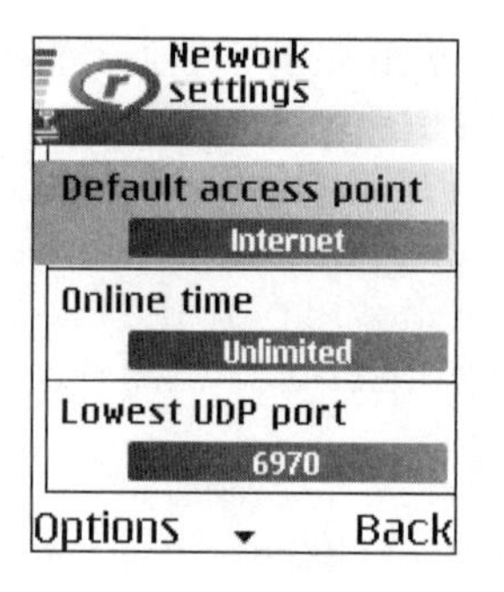

Figure 18-9

The Internet access point in RealPlayer has to be the same as that in the Web browser.

Windows Mobile Phone Windows Mobile smartphones that come with Windows Media Player software can usually show live TV received from Orb without modifications to the settings.

note *A phone that is equipped with Wi-Fi or 3G (EV-DO, UMTS, or HSDPA) connectivity is the best choice for frequently watching TV or videos. If you are subscribing to a data communication plan, make sure that the plan allows unlimited data traffic. Wi-Fi is free at your home and in many other places. It is an excellent, although power-hungry, choice for accessing Orb and other digital media on the Internet.*

Other Phones If you are unsure about the streaming capabilities of your phone, launch the Internet browser and go to test.orb.com and try it out.

Step 8: Watch TV on Your Phone

You've come a long way in this project, and now that everything is set, you can try out how your home TV looks on your phone screen.

Symbian OS/S60 Phone Launch the Web browser on your phone. Go to my.orb .com (Orb 1) or mycast.orb.com (Orb 2) and sign in to your account. Click the link for TV. Choose the TV channel you want to watch. Push the Play button or link to start streaming. You have to be patient; it may take a minute or two before the picture appears on the screen (see Figure 18-10).

tip *If the TV picture continues to be choppy after you have closed unnecessary applications on the Orb server, there is another trick you can try. You can record TV shows remotely from your phone. Log on to Orb and select a TV channel. Click the REC button on the Web page to start recording. The program is saved on your server's hard disk. When you want to watch the show, you can find it in the TV section. There is a link for Recorded TV. Play the recording. You should find it smoother than the live picture, because your server can push recorded video to the Internet faster than live TV.*

Figure 18-10

Watching TV on a phone. In this case, a Wi-Fi connection provides the fast data transmission channel for the live TV.

Windows Mobile Phone Go to the Home screen and press START. Locate Internet Explorer and launch it. Type **my.orb.com** (for Orb 1) or mycast.orb.com (for Orb 2) into the address bar and sign on to your Orb server.

Click the TV link, and choose a TV channel. Push the Play link or button. Now, you have to be patient; it may take a minute or two before you can see the picture on the screen (see Figure 18-11).

Figure 18-11

Live TV feed from your home. This phone is connected to the EDGE/GPRS network, but it can still show live TV.

tip *Orb may not provide an electronic program guide, even though the TV tuner software could show it. How can you find TV programs without an online TV guide? If your phone browser allows several browser windows, keep one window reserved for a Web page that shows the day's programming. Some browsers let you save Web pages. When you visit the Web site that provides the day's programs, save the page.*

Other Phones Any phone with an Internet browser and media player application should be able to access an Orb server. If you are unsure about your phone's media streaming capabilities, point its Web browser to the Web page test.orb.com. Follow the instructions for testing your phone.

An alternative technology to streaming live TV over the Internet is broadcasting TV to mobile devices directly from TV stations. The benefit of having broadcast technology on a mobile device is that it is not dependent on the network speed or processing power of the device. The downside is the additional cost for the device, because every unit will have to come with a built-in TV tuner. Additionally, someone has to build and pay for new, dedicated broadcast networks for mobile broadcast TV.

Three TV technologies are competing in the mobile broadcast TV market: DMB, DVB-H, and MediaFLO.

The DMB (Digital Multimedia Broadcasting) system has been designed for broadcasting data, audio, and TV to mobile devices. The DMB signal can be transmitted from satellites (S-DMB) and from terrestrial radio masts (T-DMB). The receiver on the phone can only pick up one or the other. The DMB system is already in operation in South Korea.

The DVB-H (Digital Video Broadcasting—Handheld) technology was developed from the digital TV (DVB-T) system. DVB-H is being tested in Australia, Europe, and the United States. Some of the major phone manufacturers are supporting the DVB-H standard and have announced phones that can receive DVB-H broadcasts.

MediaFLO technology has been developed for building broadcast mobile TV networks and devices. MediaFLO is the newest of the three TV technologies, and it may take some time before it reaches retail shop shelves.

Project 19

Improve Your Photos in an Imaging Application

What You'll Need:

- **A photo manager application for your phone**
- **An image editor software for your computer**
- **Cost: $20 U.S. for the phone application**
  $0–$100 U.S. for the PC application

You have just taken a beautiful picture on your camera phone and desperately want to share it with your friends—but a teenager in the background of the frame ruins everything by showing the finger. The photo is not necessarily doomed if you have an application on your phone that lets you modify the image. If the picture needs more thorough touch up, you can copy it to your computer and edit it in a dedicated imaging application.

Editing digital photos doesn't have to be an art form. Correcting minor flaws and removing unnecessary parts from a frame is a straightforward task with the right tools. In this project, you'll find out how to make small tweaks to the photos on your phone (see Figures 19-1 and 19-2) and bigger modifications to the images on your computer.

note *However, let's be realistic. It's easy to snap photos on a camera phone, but it's not that easy to edit photos using the phone's small keypad and screen. After all, not all that many people even bother to edit their digital images on a large computer screen. Nonetheless, it's fun to be able to do small tweaks on the phone, because you can instantly share the photo and don't have to wait to get access to a computer.*

Figure 19-1

Lovely shot, but a bit too dark.

Figure 19-2

The image in Figure 19-1 after it was modified on the phone.

Step 1: Install a Photo Editing Application on Your Phone

There are plenty of image viewer and photo organizer products designed for phones. I've chosen PhotoRite SP as the application for Symbian OS devices, because PhotoRite is one of the few products that let you fix images and it also comes with some fun features.

The Pictures & Videos (Photo Album) application included with every Windows Mobile phone lets you edit images. You can rotate and crop photos. These features alone can greatly enhance many pictures. Sometimes, more drastic measures are required, and then you need an application like the Resco Photo Viewer for modifying images.

Symbian OS/S60 Phone Open the Web browser on your computer and log on to Handango online store at www.handango.com. Look for the version of PhotoRite product that is compatible with your phone. Download the product on your computer.

Copy the application to your phone using a memory card, data cable, or Bluetooth. Refer to Projects 6 and 7 for detailed instructions for transferring information between your phone and your PC.

Install PhotoRite on your phone. If you copied the program to the memory card, go to the main menu, open the Tools folder, and launch File Manager. Find the application from the memory card, open it, and the installation routine will start. If you sent the PhotoRite file into your phone's Inbox, open the message and the installation process will run directly from the inbox.

After installation, you can find the application in the My Own, My Apps, Install, or Fun folder. Launch PhotoRite. When you purchase the software, you only have to type in the registration key once (see Figure 19-3). If you downloaded the trial version, select Cancel in the start-up screen, and a trial version that runs for limited period will start.

Figure 19-3
Launch PhotoRite.

Windows Mobile Phone Start the Web browser on your computer and go to www.download.com. Search for the version of the Resco Photo Viewer application that's compatible with your phone. Download and save it on the hard disk.

The Photo Viewer application is packaged in an EXE file that you must extract on your PC. Attach the data cable from your phone to the PC. Double-click the EXE file to start installation. You don't have to install the Desktop Album or PowerPoint Converter software at all if you don't need them. The Photo Viewer application that runs on the phone doesn't require them.

When the installation program asks you to complete the process on the phone, pick up your phone. Confirm that you want to install the software.

After installation, go to the Home screen and select Start. Find the Photo Viewer and launch it (see Figure 19-4).

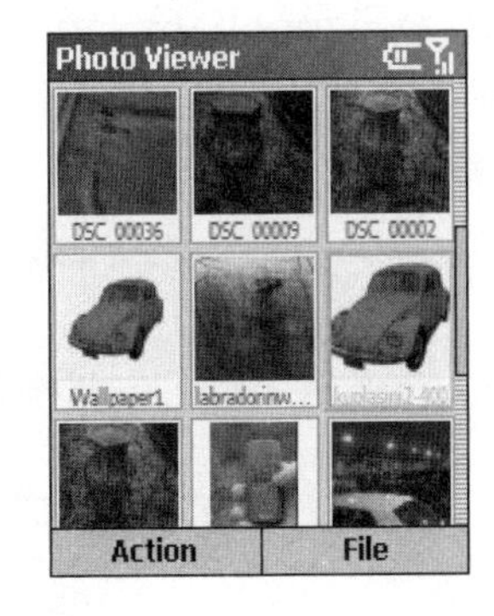

Figure 19-4
Resco Photo Viewer displays thumbnail images for easy browsing of photos.

Step 2: Edit an Image on Your Phone

Take a few photos on your camera phone, because you will need some test pictures to try out image editing.

1. Open a Picture

Symbian OS/S60 Phone When PhotoRite is running on your phone, you'll see the main menu that lets you activate the camera for taking new pictures or browsing photos stored in the phone memory or on the memory card. Navigate to your photo folder. You can view thumbnail images of pictures saved in the folder (see Figure 19-5).

Figure 19-5

Viewing a photo folder.

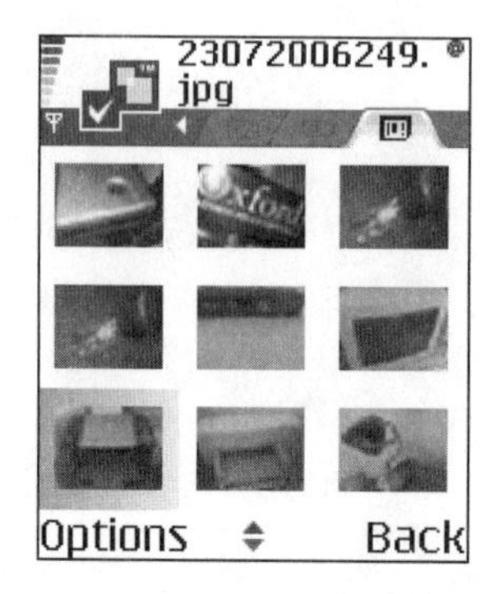

Press the selection key when you discover the photo you want to fix. The picture will open up full screen size.

Windows Mobile Phone After the Resco Photo Viewer has started up, you should see a few sample images on the screen. To add your own photos to this view, select Action | Add Pictures. Open the menu and select Browse Folder. Navigate to the folder where your photos are stored, and choose Select. You can select multiple items from the folder by scrolling the list up and down, and pressing SPACE on the keypad for each item you want. Choose Select when you are ready. The program will bring the images into the album view.

You were actually creating a photo album when you selected pictures. The Resco Photo Viewer won't remember your album unless you save it. Select File | Save | Save Album (see Figure 19-6). From now on, you can easily find this set of pictures by selecting File | Open.

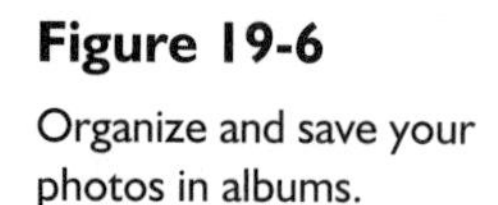

Figure 19-6

Organize and save your photos in albums.

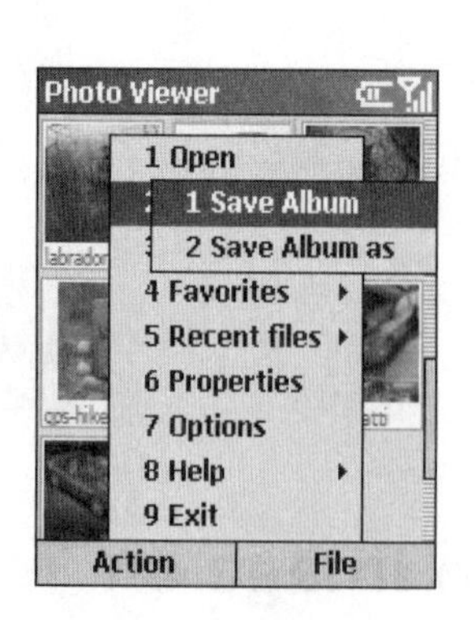

Now, highlight the picture you want to edit and press the selection key to view the full-size image.

2. Edit

Most image editing products for phones focus on fun features, such as creating nice frames, or overlaying text on the image. Explore the product you installed for other features, such as cropping and color balance as well.

Symbian OS/S60 Phone Now, when you are viewing the picture you want to edit, you can get to work. PhotoRite comes with a selection of image editing tools (see Figure 19-7).

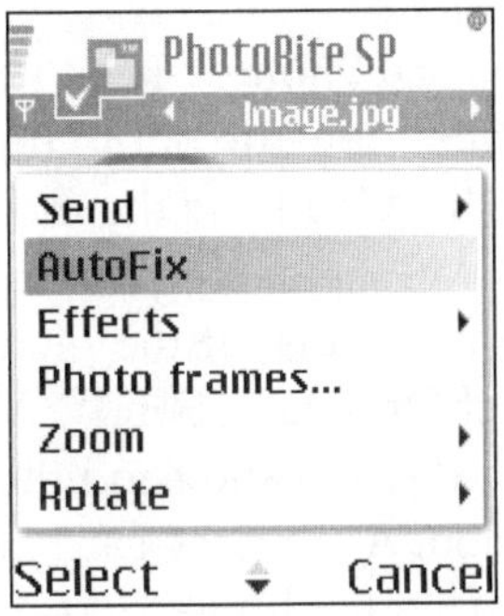

Figure 19-7

The image editing tools in PhotoRite.

Autofix can automatically fix the light and color balance of an image. This function alone can considerably improve a picture, because dark and red-tone pictures are among the most common problems in camera phone photos.

A photo that needs something special can easily be spiced up with the Soft Lens, Black & White, Sepia, Magic Mirror (a fun transformation), or Slim Up (another fun transformation) effects. Photo Frames lets you overlay graphics on a picture. Rotate turns the image 90 degrees at a time.

Don't forget to save your new picture. Open the Options menu and choose Save As to save the image.

Windows Mobile Phone When you are viewing the full-screen image of a photo in the Photo Viewer, open the Action menu. You can rotate the picture, resize the image to a smaller size, or crop unnecessary sections. These well-selected tools are easy to use even on a phone. It's worth trying them out.

In low light, camera phone photos easily get blurry and too dark. The Brightness feature in Resco Photo Viewer lets you manually adjust light and contrast (see Figure 19-8). Experiment with this feature and see how your adjustments change the image.

Figure 19-8

Adjusting brightness can save many camera phone photos.

Save the edited version of the photo. It's wise to keep the original image as well. Select File | Save As. Type a new name for the image and select Done.

3. Share the Edited Photo

The whole point of editing an image is to be able to show an exciting picture of something fun or important to friends, family, and possibly the world. You can share your photo by e-mail or by posting it on a photo-sharing site, or you may copy it to your computer for further processing.

Symbian OS/S60 Phone The PhotoRite application allows you to share a picture by e-mail or MMS. View the photo full screen and select Options | Send | Via E-mail or Via Multimedia. Compose the message and send it. There's more information about e-mail and MMS messaging in Project 10.

You can also transmit the picture to your computer via Bluetooth. View the image and select Options | Send | Via Bluetooth. Select the target device from the list of discovered devices and send the picture. Project 3 discusses more about Bluetooth communication. A memory card and a data cable are other methods for transferring pictures from your phone to your computer. Read Project 2 for more information about using a memory card as a transfer medium.

Windows Mobile Phone You can send the edited photo from the Resco Photo Viewer by e-mail or MMS. Open the picture and select File | Send | E-mail or MMS. Compose your message and send it. Project 10 provides more information about e-mail and MMS communication.

If both your phone and computer have Bluetooth connectivity, you can transmit the picture over Bluetooth as well. Open the image and select File | Send | Bluetooth. Select the target device from the list and send the picture. Project 3 discusses more about Bluetooth communication. A memory card or a data cable are reliable methods for transferring pictures from your phone to your computer. View Project 2 for more information about using memory cards.

Step 3: Install a Photo Editor Application on Your Computer

Easy-to-use and affordable image editing applications for PCs became available when digital cameras became popular. We are going to use Picasa 2 in this project, but there are many other products on the market. Picasa 2 is a free photo organizer application with some image editing features.

note *To name a few products, ACDsee is a powerful photo management package that provides image editing features. Photoshop Elements and Paint Shop Pro are advanced photo editor applications that require time to learn and master, but you can do practically anything with them.*

Launch the Web browser on your computer and go to www.picasa.com. Follow the links to download and save the product on the hard disk. Follow the instructions to install the program.

Once the installation is finished, you have to tell Picase where it can find your pictures. Launch Picasa. Open the File menu and select Folder Manager. You'll see a folder tree (see Figure 19-9). Highlight a folder and click Scan Once or Scan Always for the folders you want Picasa to catalog for you.

Figure 19-9

Let Picasa know where your photo folders are.

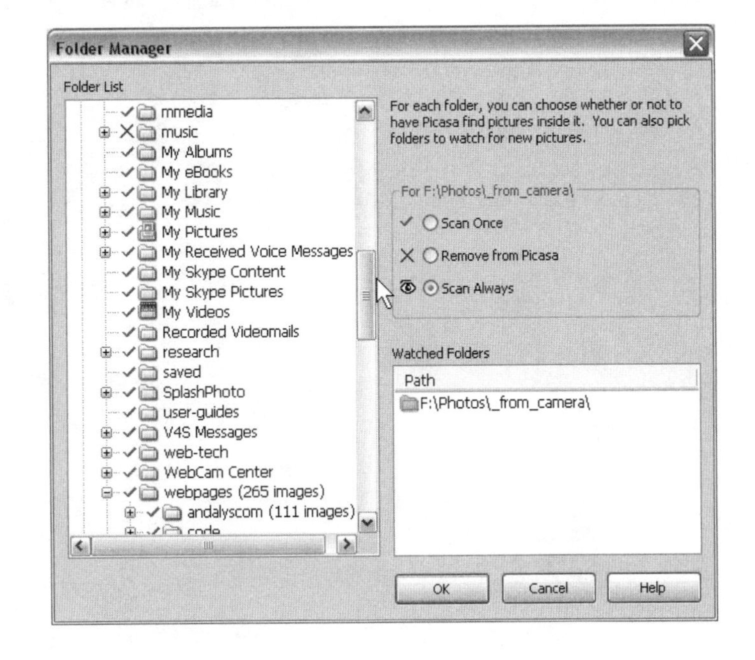

Step 4: Edit a Picture on Your Computer

Image editing is not black magic performed by highly-respected professional anymore. With a little practice, you can easily learn to improve your camera phone photos. In any case, it is good to have back-up of your images, for instance on a CD, before applying your magic into them. Project 16 discusses about backing up pictures.

1. View Folders and Thumbnail Images

Picasa lists photo folders and subfolders in a frame on the left side of the program window. Click a folder and you can view thumbnails of photos (see Figure 19-10). When you highlight a picture and click Hold, Picasa places it in the Picture Tray down left. You can collect a number of photos into the tray, which is helpful when you are looking for photos for a specific purpose. The slider down on the right lets you adjust the size of thumbnail images.

2. Edit

Double-click a thumbnail to view a photo. When the image opens up, a set of tools appears on the left (see Figure 19-11). Even if you are unsure which tool gives you the result you are looking for, simply try different options. For example, if you adjust the Fill Light, but the picture only gets worse, Undo the action.

Figure 19-10

Browse photos in Picasa.

Figure 19-11

Applying Auto-Contrast increased depth for this image.

Click the Tuning tab on the left for more advanced controls that let you adjust light, highlights, shadows, and color in the photo.

The Effects tab provides you with some artistic tools. You can, for example, turn the picture into a black-and-white photo, make it glow, or soften or tint it.

When you are done with your edits, click the Back To Library button in the top-left corner of the Picasa window.

3. Save and Find Your Original Images

Picasa automatically saves your modifications to the image. It does, however, save the original picture as well. Picasa creates a folder titled Originals under the folder where your edited photo is located. You won't Originals folder in Picasa, but you have to launch Windows Explorer and open the folder to get the original image.

Project 20

Direct Your Own Home Movies

What You'll Need:

- **Video editing application for your computer**
- **Cost: $0–$100 U.S.**

Since practically every camera phone comes with a built-in video camera, you might expect to see an endless show of home videos when you visit a friend who also happens to be an avid photographer. For one reason or another, it doesn't seem to be so. Maybe camera phone flicks are not yet perceived as home videos, or maybe clips are simply being posted on video-sharing sites without worrying too much about the way they look. In this project, you'll learn an easy way to edit and mix your camera phone videos into movies that you are proud to present to your audience at home or on the Internet.

Even if you have tried editing digital still photos, but felt that it wasn't something you would like to master, you should try video editing. The tools that we will use are extremely easy to use. Adobe Photoshop Album Starter Edition and Windows Movie Maker are simple products in a sense that they are easy to use but they also have their limitations. Later, if you want to upgrade to more advanced tools, you can apply what you've learned from these entry-level applications. We are not going to mix a sequel to *Star Wars,* but we'll put together a snappy home movie by cutting and pasting sections from the material you have captured.

The whole video editing process consists of five major steps:

1. Capture video clips on your camera phone.
2. Copy the material to your home PC.
3. Install a video editing product on your PC.
4. Put together a movie from multiple video clips.
5. Show the finished movie on a large-screen TV.

Let's start by setting up your camera to be ready for a video capture session.

Step 1: Capture Video on Your Camera Phone

Before diving into the world of (home) movie stars, you might want to refresh your memory with the tips in Project 4. Then, prepare your camera phone for some video action.

1. Set the Image Resolution

Before you set off to shoot movies on your camera phone, make sure the video is set to its highest resolution. The low video resolution is intended for MMS for minimizing the size of a message sent to another phone. Because you are going to copy the captured videos to your computer, or upload them to the Internet, you should choose the highest video resolution. The difference between resolutions may not be obvious on the phone screen, but when you see low- and high-resolution videos side by side on a computer screen, you'll know what you want (see Figure 20-1).

Figure 20-1

A 3GP video in 176×144 resolution (right) and an MP4 video in 320×240 resolution (left). Both clips are scaled up to double of their original size using the controls in the player applications. The picture was taken from videos clips displayed on an LCD screen.

note *The number of megapixels manufacturers advertise won't tell you anything about the camera phone's video recording capability. For example, a one-megapixel camera phone may capture low-quality video clips in 176×144 pixel resolution. If the video is not properly shot, it can be difficult to see what's going on in the movie. On the other hand, another one- or two-megapixel camera phone may be capable of recording movies in 320×240 resolution and of a quality that is a pleasure to watch on a computer monitor or TV screen.*

2. Choose the Best-Quality Video Format

In addition to the 3GP video format, some phones can record AVI or MP4 (MPEG4) movies. These digital video formats save images in higher quality than 3GP. Check your phone's video settings for AVI or MP4 and choose either one if available.

Activate the camera and change to video mode. Open the menu and look for video quality settings. Select the highest video quality.

3. Define the Storage Device

Always save the captured video clips on a memory card. You get more memory space for your recordings, and it is easy to copy the material to a computer.

Set the memory card as the default storage in the camera settings. Activate the camera, open the menu and change the default storage in the settings to Storage Card or Memory Card.

4. Capture

Capture as much material as you can. For further advice about shooting video on your camera phone, read Project 4. Once you learn to use the video editing tools, it'll be easy to cut and paste movies from the material you have.

Step 2: Copy Camera Phone Movies to Your PC

Movie files, except for the 3GP videos, tend to be large in size. The fastest and most reliable way to transfer videos from your camera phone to your computer is to use a memory card or data cable.

Remove the memory card from your phone and insert it into a card reader. Using Windows Explorer, copy the video file to a folder on the computer where you can find it later. Read Project 2 for more information about transferring data between your phone and PC.

Step 3: Install a Video Editing Application on Your PC

The video editing tools presented in this project don't require excessive memory space or processing power from your PC. The learning curve is also considerably shorter than in full-blown video editing tools that are available, for example, from Adobe and Pinnacle.

1. Choose Your Tool

Practically all camera phones that can capture video can save the material in 3GP format. Adobe Photoshop Album Starter Edition is an application that can edit 3GP and MP4 digital films. Adobe has made this product, which is primarily intended for organizing photos, available for free.

If your phone can capture videos in other than 3GP format, you might consider using Windows Movie Maker as your video editor. For example, phones powered by the Windows Mobile 5 operating system can save movies in AVI format. Windows Movie Maker can edit AVI videos, among other common video formats used on computers. The product comes with Windows XP Service Pack 2, but it is also available as a free download.

tip *If you want to edit a video while it is still on your phone, the choices are limited, but there is something you can do. An application called Muvee (www.muvee.com) can create a new movie from video clips that you have recorded and saved on your phone. The product takes segments from the clips and overlays graphics and music over them creating fun video clips.*

2. Install the Video Editing Application

Before paying hundreds of dollars for an advanced video editing product, you can try out the following free products in order to see if video editing is something for you.

Adobe Photoshop Album Starter Edition Go to the Web page www.adobe.com/products/photoshopalbum/starter.html. Download the product on your computer's hard disk. The downloaded item is a ZIP file. Open the file and double-click the EXE file inside the ZIP file to initiate installation. Follow the instructions to complete the installation.

Windows Movie Maker If your computer runs on Windows XP, you may already have Movie Maker installed. Open the Start menu, select All Programs, and view the application titles. Usually, the product is located in the main menu, but check other folders as well. If you can't find it, and you have Service Pack 2 (SP2) installed, you can download the application from this page: www.microsoft.com/windowsxp/downloads/updates/moviemaker2.mspx. Follow the links to download and install the application.

Step 3: Edit Your Video

It is difficult to shoot a perfect video clip with one take. Be bold and experiment with your material. Cutting, pasting and reordering the clips can easily turn a boring camera phone video into a decent home movie.

1. Open a Video Clip

Adobe Photoshop Album Starter Edition Because Photoshop Album is a product for organizing, searching, and viewing images, the first thing you have to do is to tell the program where it can find your videos and photos. Launch the application. Select File | Get Photos | From Favorite Folder. Navigate to the folder where your videos are stored.

The Photoshop Album will collect images from the folder and organize them into the main screen where you should see thumbnail images of your photos and videos. Video clips are identified by a small film symbol in the corner of the thumbnail image (see Figure 20-2). Double-click a video to play it. Although you can view many types of digital videos in the program, it only lets you edit 3GP and MP4 files.

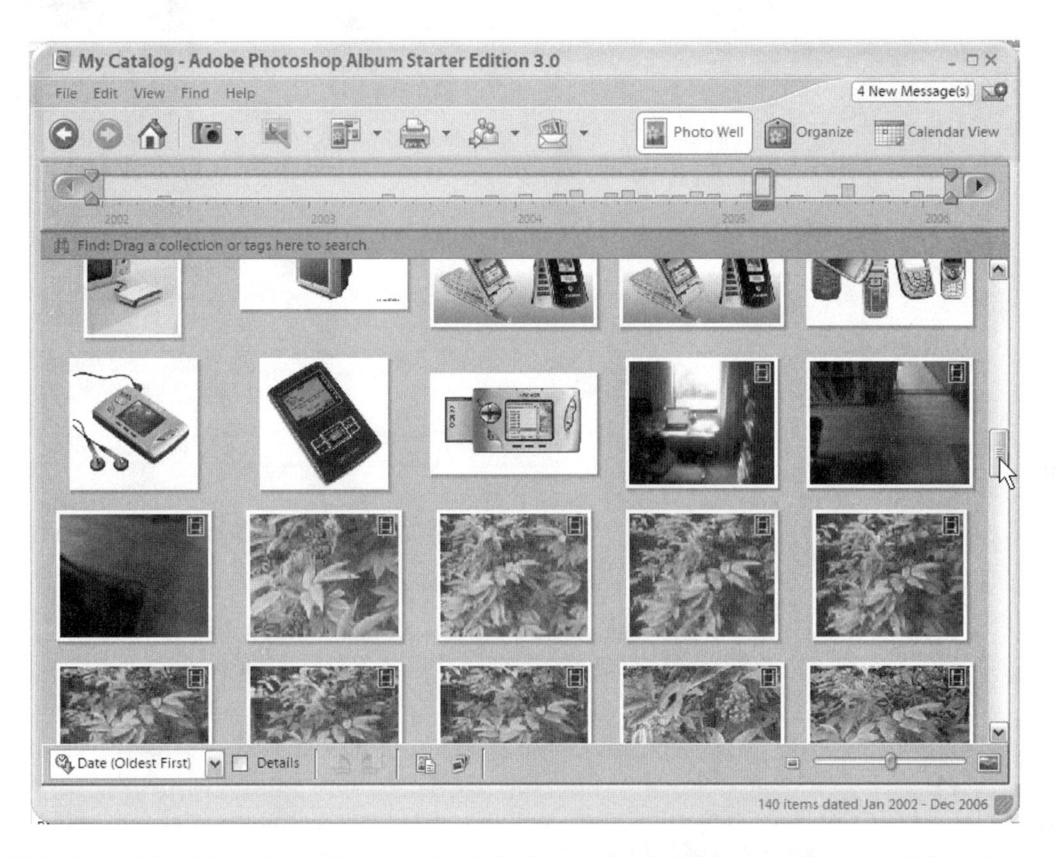

Figure 20-2
Photos and videos are neatly organized in the Adobe Photoshop Album Starter Edition.

Windows Movie Maker Launch the Windows Movie Maker application. When the program opens up, you should see a pane on the left titled Movie Tasks. Click on the Import Video link. Find the folder where you copied your videos and select a movie you want to edit. In the video player pane on the right, click the Play button to watch your raw material (see Figure 20-3).

2. Create a Single-Segment Movie

All movies, long and short, zero-budget and big-budget, are composed of segments that make up the story. Your task is to identify the interesting segments from your videos that contribute to the story you want to tell.

Adobe Photoshop Album Starter Edition When you are viewing your photo and video collection in the main screen of the program, right-click a thumbnail of a video. Select Edit Mobile Movie from the pop-up menu. A new video editing window will open up.

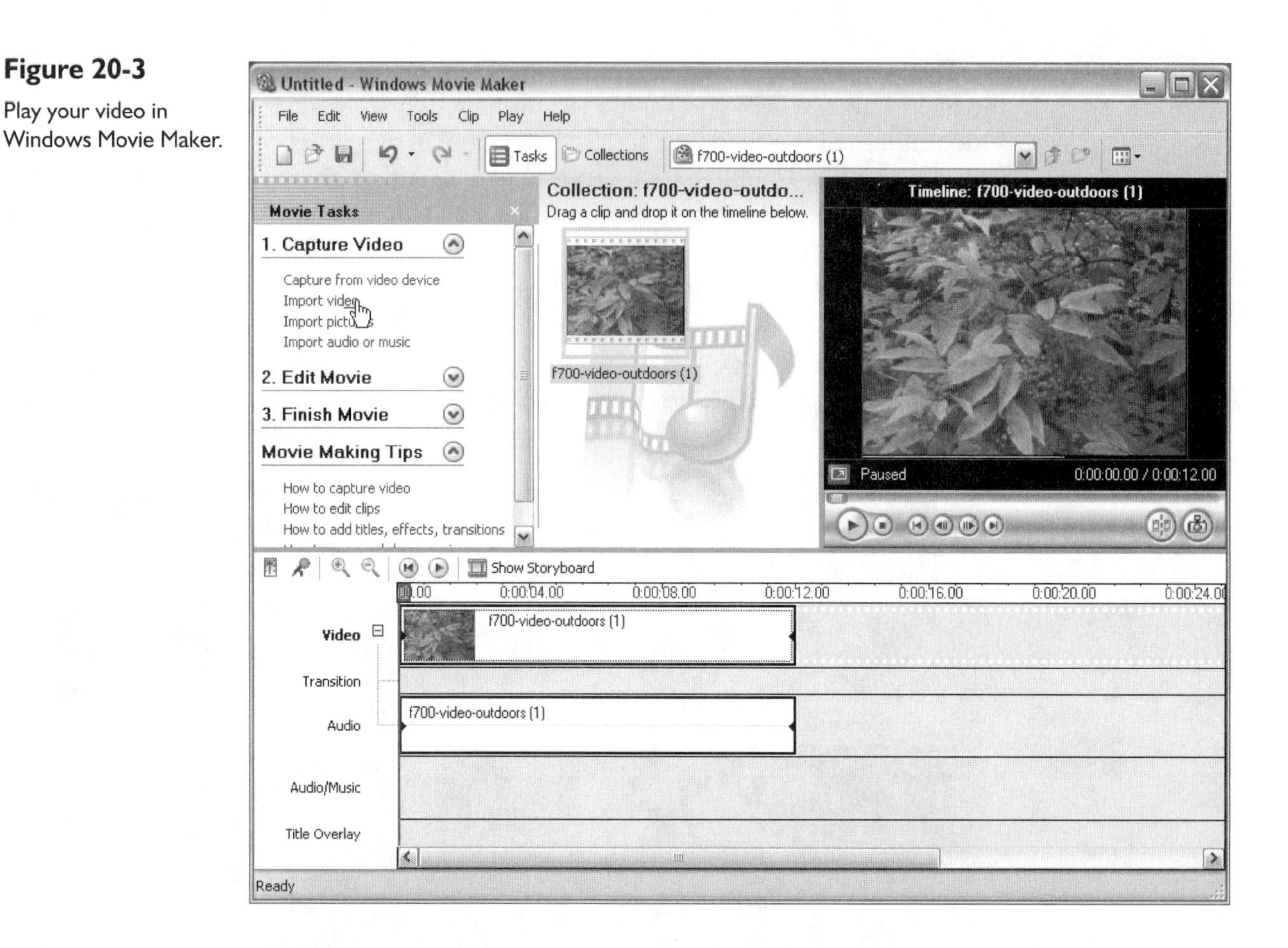

Figure 20-3

Play your video in Windows Movie Maker.

The top half of the program window is dedicated to viewing the video under work. Play the video by clicking the Play button below the video pane. The software has cleverly sliced the video into brief segments that you can see side by side at the lower half of the program window. You should see the red Timeline Marker moving across the segments when you play the video.

You are going to extract a segment from the video. Grab the red Timeline Marker with your mouse and move it to the beginning of the segment you want to preserve. Mark the start point by clicking the Set Mark In button just above the video frames. Then, move the Timeline Marker to the end point of the segment. Mark this point by clicking the Set Mark Out button (see Figure 20-4).

Click the Trim Selection button to clean the rest of the video. You have just created a movie that consists of one segment only. If it delivers the story you want, that's it—you're done.

Windows Movie Maker When you have imported your video into the center pane of the Windows Movie Maker window, select it and view it in the player pane on the right.

View your original video recording by clicking the Play button in the pane on the right. Identify the beginning of the segment that you want to preserve. Stop the video at the start point. Click the Split button that's under the video picture. The video you

Figure 20-4

Select a segment from your video.

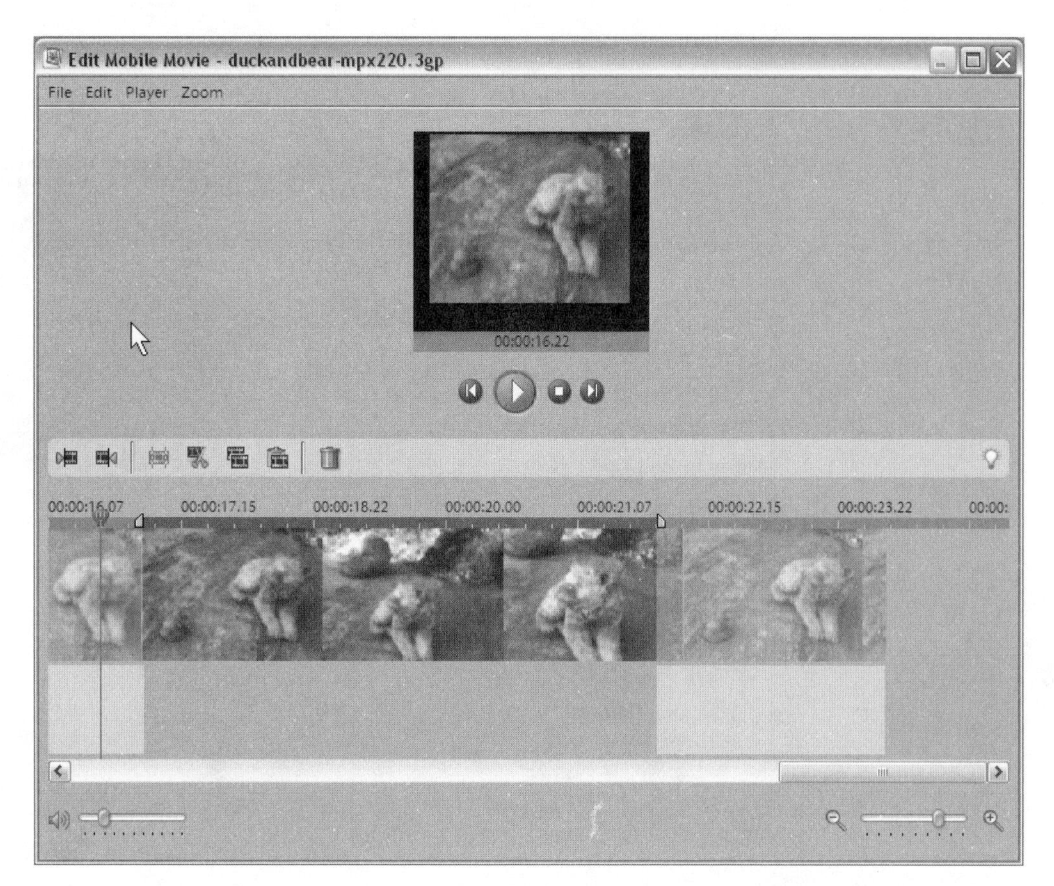

are currently viewing is split into two pieces. You'll notice that there are now two video thumbnails in the center pane (see Figure 20-5). Select the second thumbnail, and play it in the video pane until you find the end of the segment you want to keep. Again, click the Split button.

Let's assume that you want to save the middle section from the original video. Grab its thumbnail image with the mouse and drag and drop the clip from the center pane into the Storyboard area. The Storyboard/Timeline area occupies the entire lower half section of the Windows Movie Maker screen. The area has two functions: the Storyboard lets you rearrange segments, and the Timeline lets you add transitions, music, and text to the videos. You can switch between the functions by pushing the text beside the tool icons.

Highlight the clip you dropped into the Storyboard and click Play to view it. If it is what you wanted, that's it—you have just created a snappy home video.

3. Create a Movie from Multiple Segments

Often, there is more than one good sequence in a captured clip. Perhaps you want to preserve multiple segments or change the order of the segments for the final movie. It's easy to do.

Figure 20-5

Split your original material into segments.

Adobe Photoshop Album Starter Edition Mark the segment in the video that you want to keep. Drag the Timeline Marker and use the Set Mark In and Set Mark Out features to make the selection. Click the Cut Selection button in the toolbar above the frames to save the section into the clipboard.

Move the Timeline Marker to the new position where you want to move the segment. Click the Paste button on the toolbar above the frames. The saved sequence is pasted just before the marker (see Figure 20-6).

Remove the sequences that you don't need in the final movie by marking the segments and trimming or cutting them. If you happen to make a mistake, you can track back by opening the Edit menu and selecting Undo.

tip *Adobe Photoshop Album Starter Edition doesn't let you import and merge segments from other video files to the movie you are currently working on. If you want to edit segments from multiple files and combine them into a single movie, you can convert your 3GP or MP4 clips to, for example, AVI format. Then, you can continue working in another video editing application. For example, Video Converter 2005 application can do this conversion for you.*

Windows Movie Maker In order to put together a movie that consists of multiple segments, you need to do some work in the Storyboard. Start by bringing the segments you have extracted from the original video into the Storyboard. Once the clips

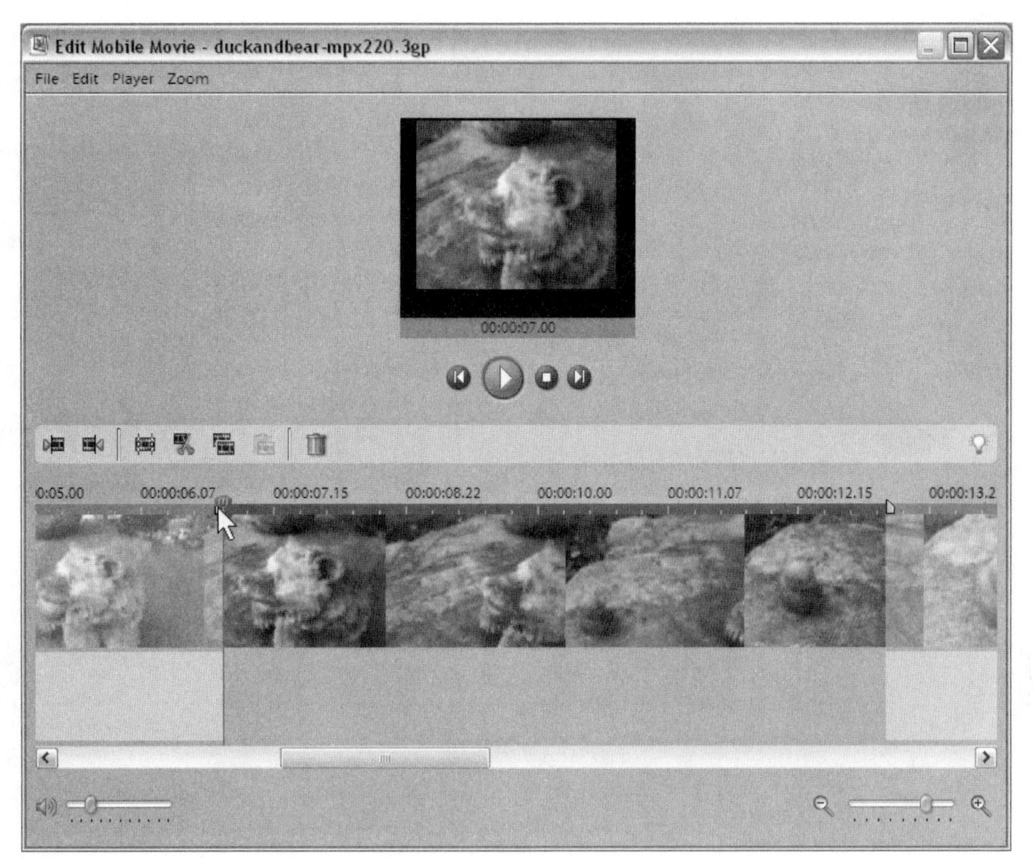

Figure 20-6
Cut and paste segments to create a new movie.

are in the Storyboard, you can shuffle them into any order you like. Drag and drop segments into free slots in the Storyboard.

You can also merge segments from other video files. Bring new material into your current workspace with the Import Video feature. You can work with the new material just like you did with the first movie. Split segments, drop them into the Storyboard, and arrange them to create a new movie (see Figure 20-7).

4. Save Your New Movie

The more you work with videos, the more versions of the same work you probably will save. It helps you watch different versions and decide which one is the best for public viewing. For the sake of this project, let's save the final cut only.

Adobe Photoshop Album Starter Edition Once you are happy with your new movie, open the File menu and choose Save. The Photoshop Album will automatically save the video in the same folder where the original file is. The program also adds the text **_edited** to the end of the file name so that the original file stays intact.

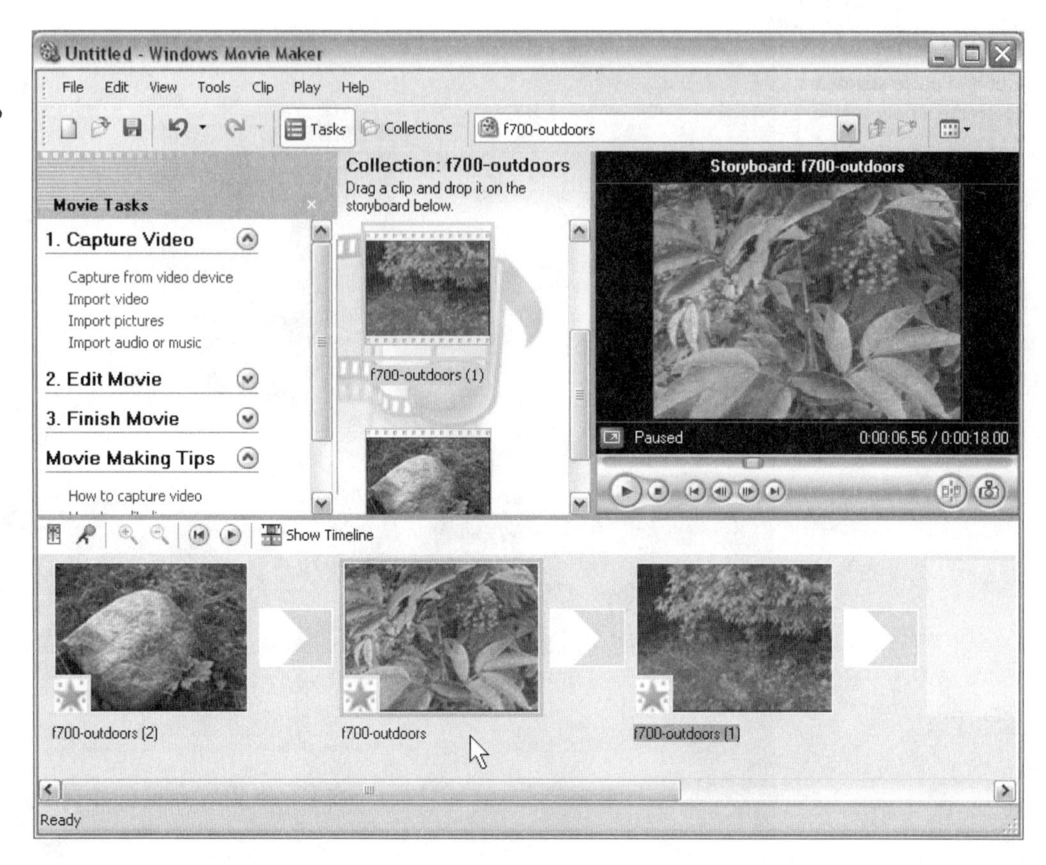

Figure 20-7

Combine segments from multiple videos to create a new movie.

Windows Movie Maker You have two options for saving a video in Windows Movie Maker: save the whole project that you are working on, or only save the final movie.

Save the final movie that you created into the Storyboard. Select File | Save Movie File | My Computer and choose the folder where you want save the movie. Then select Best Quality for Playback on My Computer and wait for the file to be saved.

tip *You can optimize the movie resolution and quality for viewing on different devices. When saving the movie, there is a link titled Show More Options under the choice Best Quality for Playback on My Computer. Click the link and you can select Other Settings. Open the drop-down menu, and you'll see a list where you can choose the option closest to your viewing device.*

Step 4: Show the Video

It's fun to watch videos on the computer screen, but if you intend to have a larger audience for your flick, you might want to show the film to guests and family on a large-screen TV in your living room. Friends and relatives who don't live nearby may enjoy your home videos on a video-sharing site on the Internet.

TV Set If you want to play your camera phone videos on your TV, you have to figure out a way to hook up your computer to the TV. You will need an S-Video, RCA, or other type of cable. Test the setup for making sure that the video picture looks fine on the TV. The picture won't be crystal clear, but it should be worth watching. Take a look at Figure 20-1. It gives you some idea of the differences in image quality and size. You can find more details for connecting a PC to a TV in Project 15.

Internet If you have ever watched home videos uploaded to Blip.tv, Guba, Revver, YouTube, or any other popular video-sharing site, you realize that it shouldn't be too difficult to make better movies than many home videos already uploaded to the Internet, especially now that you have the tools and skills to edit your own home movies. Sign up to a service you like and upload your video to your online video album.

tip

You can also upload videos to the video-sharing sites from your camera phone. The easiest technique is to use e-mail. You need an e-mail application for your phone that lets you send large attachments. Project 13 explains the techniques for posting videos from your phone to blip.tv, YouTube, and other online video albums. Also, the ShoZu software can automatically upload your camera phones videos to the Internet.

Project 21

Post Your Photos and Thoughts on a Blog

What You'll Need:

- **An account with a blog system**
- **A blog application for your phone**
- **Cost: $0–$10 U.S. per month for blog service account**

Blogs are personal journals, or diaries, that anyone can publish on the Internet under his or her own name or using an alias. You simply sign up to a blog service and start writing. Many blog systems are also capable of publishing pictures. If you develop into an active blogger, who always can't wait to get to the computer keyboard, you can post new entries from your phone as well.

You have three options for blogging from your phone: e-mail, dedicated application, or Web browser.

Bloggers who like to express themselves in writing should find e-mail to be the most flexible way to blog on the go. This technique works with all blog systems that can receive new posts by e-mail. We'll go through the steps for posting to a blog hosted by the TypePad system, but the process is similar for other blog services that accept new entries by e-mail.

note *If blogging on a small numeric keypad doesn't sound like a good idea, there are products that feature a full alphabetic keyboard but still manage to stay small in size. For example, RIM Blackberry products, Motorola Q, Nokia E61 (see Figure 21-1) and Treo 600- and 700-series are popular devices for typing e-mail messages and notes.*

Bloggers who primarily post photos may download and install a dedicated photo upload application on their phones. An application like this automatically posts new pictures on your album or creates a new entry in your blog. You'll find out how to use ShoZu for Buzznet photo blog and TypePad Mobile for posting on your TypePad blog.

Figure 21-1

A variety of small, palm-size devices with a Qwerty keypad are available from major manufacturers.

If you have already been blogging, you may be used to doing it in a browser application on your computer. Advanced browser software on your phone lets you access your blog as well. The drawback is that usually blog systems have been designed for computer screens. Only a few blog systems provide pages designed for small screens. You'll learn how this technique works with a blog hosted by WordPress.

Step 1: Blog via E-mail

First, let's see what you have to do in order to post a new entry to your blog from your mobile device using e-mail.

1. Make Sure You Can Send E-mail from Your Phone

Compose a new e-mail message on your phone and send it your own e-mail address. Check your inbox. If you received the message, you are ready to start mobile blogging. If you didn't, view Project 10 for help on how to set up and use e-mail on your phone.

2. Get Your E-mail Address for Posting to TypePad

Log on to your TypePad account using the Internet browser on your computer. If you don't have a blog account yet, get it now. Go to www.typepad.com, click the Sign Up link, and follow the instructions. If you are using another blog system that accepts new posts by e-mail, you should be able to follow this guide as well. The fundamentals for posting via e-mail are similar in other services.

Once you are signed in to TypePad, click the Control Panel tab at the top of the window. A new set of tabs will be displayed. Select the Profile tab and click Mobile Settings below the tab.

note *TypePad provides a number of options for posting via e-mail. We'll use the secret address option, because it is safe enough without being too difficult to use. You get a unique, secret e-mail address for posting, and TypePad also checks the e-mail addresses from where the new posts are coming. You can post from your mobile device and from your computer, even if you have different e-mail accounts for them.*

Scroll down the TypePad Mobile Settings window until you find "Indicate the email addresses you will be using" and "Select how you would like the system to verify your messages" (see Figure 21-2). Enter the e-mail addresses you want to use for posting on your blog. For instance, if you have a dedicated mobile e-mail address, you should enter it here. You don't have to fill in the e-mail address you already have in your TypePad profile. Write down the secret e-mail address for the next step.

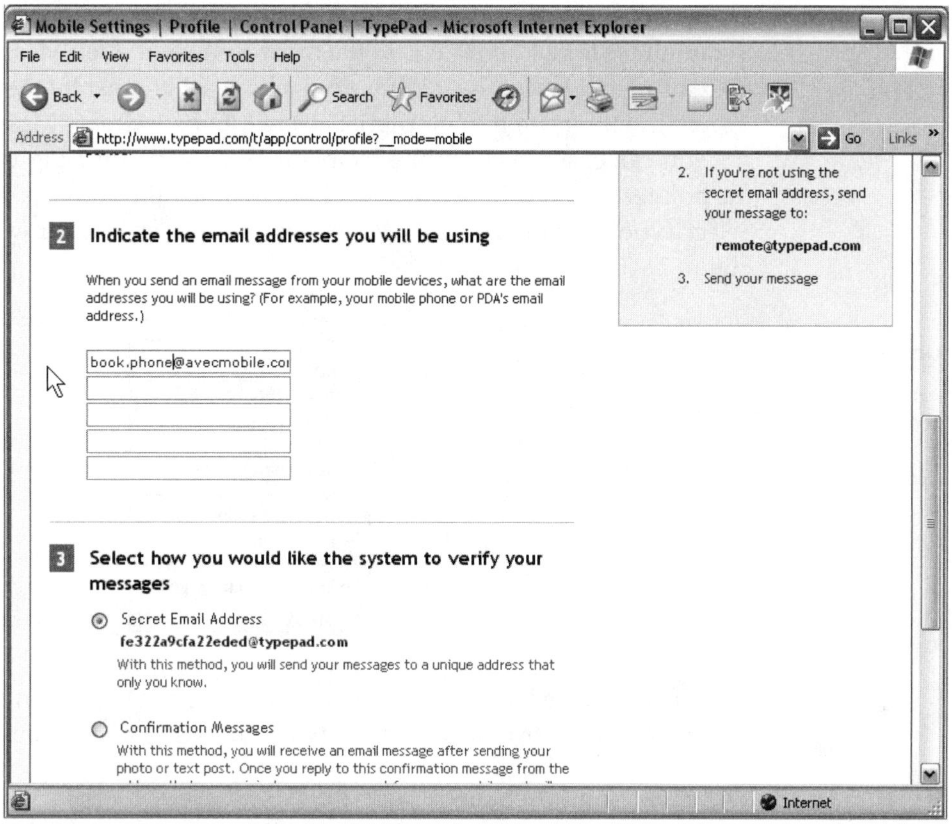

Figure 21-2

Fill in the e-mail addresses you will be using for blogging and write down your secret address.

3. Save the TypePad E-mail Address on Your Phone

The TypePad e-mail address is cryptic on purpose. Because you will probably need the address more than once, it's worth saving it in the phone's contact database.

Symbian OS/S60 Phone Go to the main menu and select Contacts | Options | New Contact. Enter TypePad into the first name. Scroll down until you find E-mail. Fill in the secret address that TypePad provided for you.

Windows Mobile Phone In the Home screen, select Contacts | New Contact. Enter TypePad into the First Name, or something else that describes your blog's address for you. Scroll down until you find E-mail. Fill in the secret address that TypePad reserved for you.

Other Phones Add your personal blog's TypePad e-mail address to the phone's contacts or address book.

4. Write and Post a New Entry to Your Blog

tip *You can write your blog entry without a network connection. Only when you want to post do you have to be in the network coverage.*

Symbian OS/S60 Phone Go to the main menu and open Messaging. Open an e-mail account you have configured for your phone (this must be the account you saved in TypePad settings on the Internet). Select Options | Create Message | E-mail.

You can easily pick up the recipient address by pushing the selection key when the pointer is in the To: field. The phone will list the contacts with e-mail addresses. Select TypePad from contacts. Write your blog entry as you would normally do (see Figure 21-3).

Figure 21-3
Update your blog from the phone.

When you are happy with the blog entry, open Options and select Send. TypePad soon confirms your new blog entry with a message to your inbox.

Windows Mobile Phone Go to the Home screen, press START, and launch Messaging. Select the e-mail account you saved in the TypePad's Mobile Settings. Select New. When the pointer is in the To: field, you can push the selection key to view contacts that have e-mail addresses. Pick up TypePad from the list. Type your blog entry (see Figure 21-4).

Figure 21-4
Write and save a new blog entry on your phone.

Select Send when you are done. Open the menu and select Send/Receive (see Figure 21-5). Depending on your e-mail settings, you may have to type your e-mail user name and password. TypePad will send you a confirmation message for the new entry on your blog.

Figure 21-5

Send the new entry to your blog.

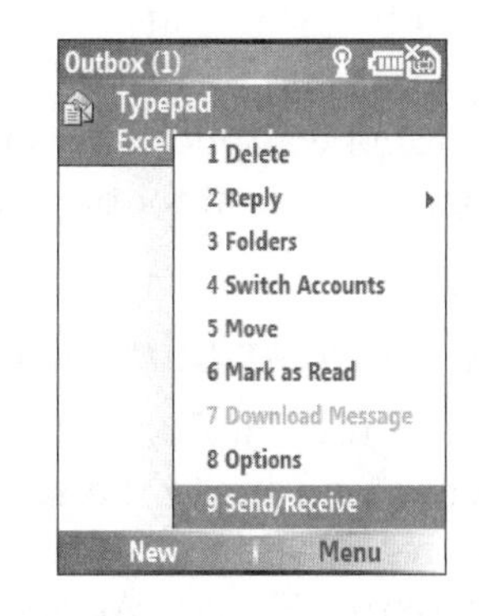

Other Phones Compose a new e-mail message. Select TypePad as the recipient. Write your new post and send it.

5. View Your Blog

When you get access to a computer, take a look at your blog (see Figure 21-6). Your personal page is your_user_id.typepad.com, where the first part before the dot is the user name you chose for your account when you signed up.

Figure 21-6

Readers won't see any difference between ordinary blog entries and posts from a phone.

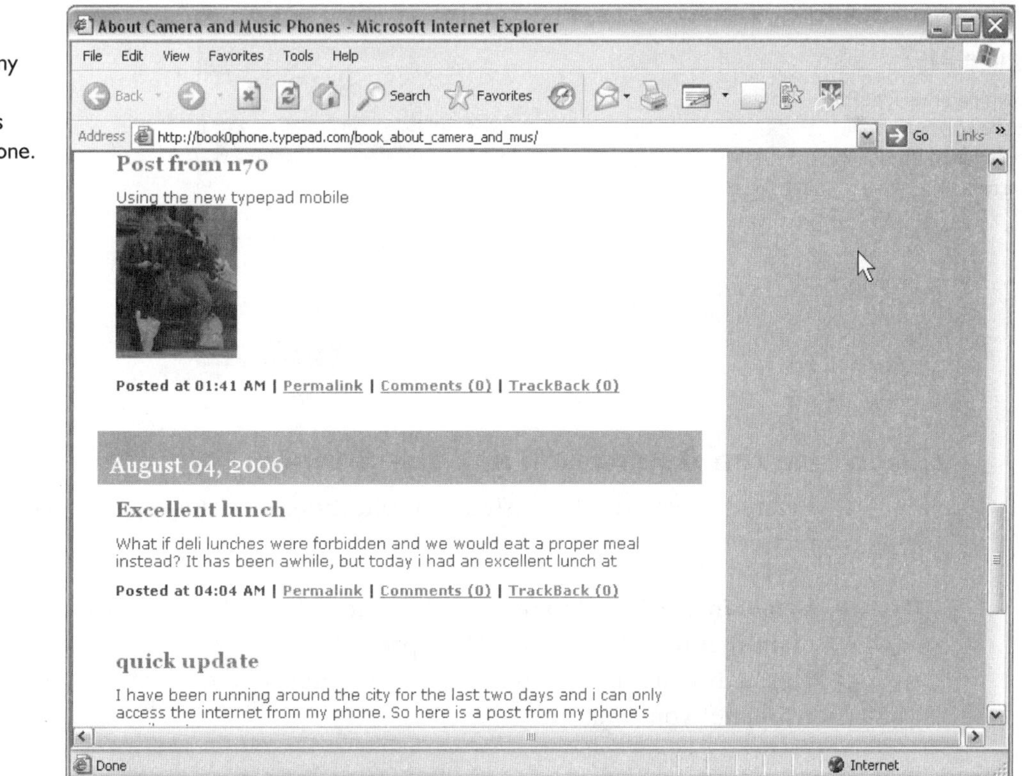

Step 2: Blog Using an Application on Your Phone

Dedicated blog applications that run on a phone are particularly useful for frequent photo bloggers. ShoZu application can post your photos and descriptions to several popular photo blogs and also to other Web sites. In this project, we'll use ShoZu for posting pictures on Buzznet.

TypePad Mobile application is handy for publishing text and photo entries on a TypePad blog. We'll use it for blogging on a Windows Mobile smartphone.

1. Download Your Blog Application

If you intend to use ShoZu, go to the ShoZu home page at www.shozu.com using your computer. Sign up to the service. Make sure your phone is compatible with ShoZu before registration. Also, if you haven't signed up to a photo blog service, do it before registering to ShoZu. We are going to use Buzznet (www.buzznet.com) in this project. Follow the instructions on the ShoZu Web page to download and install the application. I downloaded ShoZu into a Symbian OS/S60 device for this project.

If you are a TypePad blogger, go to get.typepad.com and ensure your phone is compatible with the software. The fastest way to install the application is to visit get.typepad.com in your phone browser (see Figure 21-7). Select the application version that is compatible with your phone and start the download. Installation should begin automatically once the download is ready. We are going to run TypePad Mobile on a Windows Mobile device.

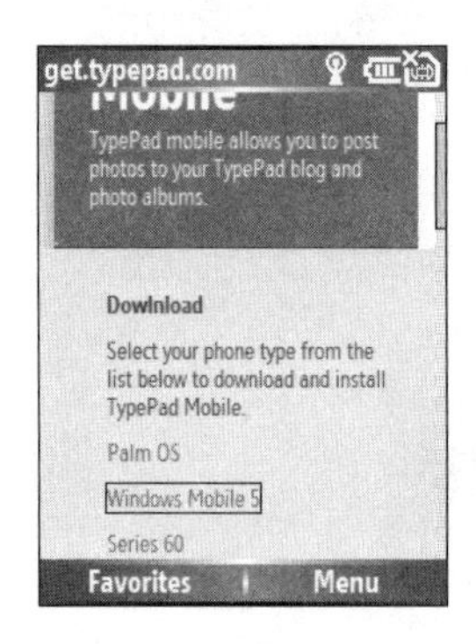

Figure 21-7 Download the TypePad Mobile to your phone directly from the Internet.

2. Activate the Application and the Service

Before you can run the installed application, you must specify which blog you want to access.

Symbian OS/S60 Phone You have to activate ShoZu after installation. Go to the main menu and open the My Apps folder (or My Own, Fun, or Install). Launch ShoZu. It will ask for you to log on. You must enter your ShoZu user name and password, not your Buzznet user name.

Windows Mobile Phone Push START in the Home screen. Find TypePad and launch it. You have to log on to the blog service.

Other Phones Follow the instructions for the application you downloaded.

3. Post a Text and Photo Entry to Your Blog

A really useful feature both in ShoZu and TypePad Mobile is the possibility to write something, edit it, add a photo, change your mind, and keep it all saved on the phone. Once everything is finally, completely ready, you can post both the text and photo to your blog at one go.

Symbian OS/S60 Phone Activate the camera on your phone. Take a picture. ShoZu will wake up and ask for your confirmation for uploading the image to Buzznet. Don't upload yet, but select More Options and Add Details (see Figure 21-8). Now, you can enter a title for the blog entry and write your thoughts in the Description box.

Figure 21-8

Accompany your photo with some thoughts.

When you have added details, open Options and select Save to Buzznet. ShoZu will post the photo along with the text.

Windows Mobile Phone When the TypePad application is running, select Tools | New Post. Type your entry and attach a picture. You can add the picture when you open the menu and select Add Photo (see Figure 21-9).

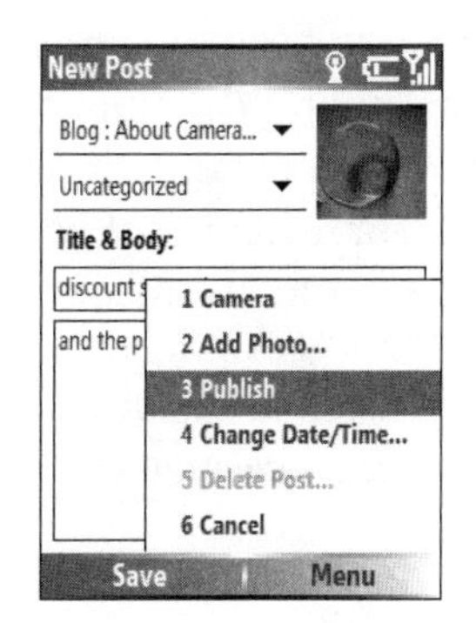

Figure 21-9

A new entry for your blog.

When you are happy with your new entry, open the menu and select Publish. Your new blog post will soon appear on the Internet.

Other Phones Follow the described guidelines for ShoZu or TypePad Mobile.

4. View Your Blog

When you have access to a computer, browse your blog (see Figures 21-6 and 21-10).

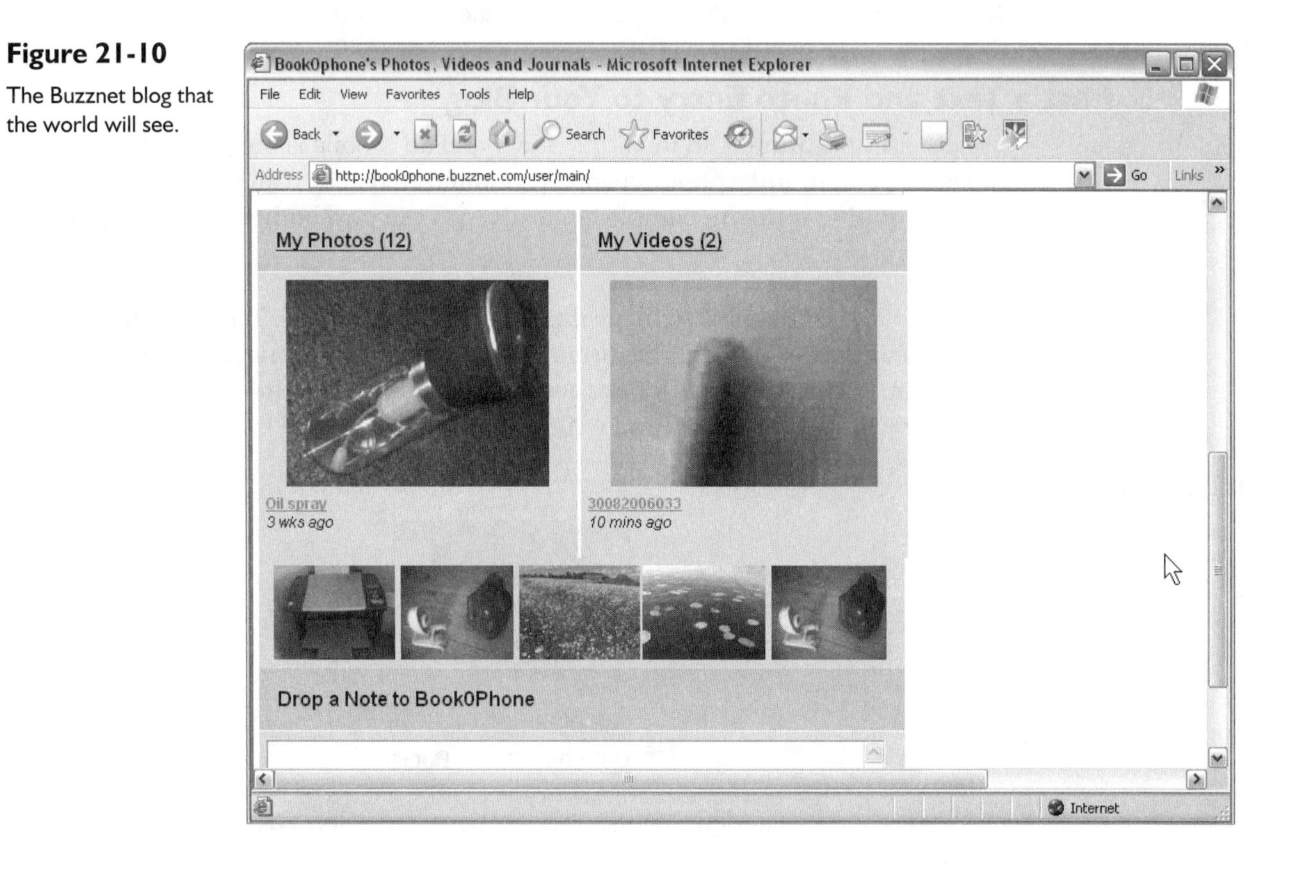

Figure 21-10
The Buzznet blog that the world will see.

Step 3: Access Your Blog on Your Phone's Internet Browser

If you have already been blogging, you are probably used to doing it in a Web browser on your computer. Why not use the browser application on your phone as well? It is possible, but somewhat tedious because of the small screen. Phone models specifically designed for e-mail that come with a large keypad and large screen are good choices if you want to use the mobile browser for blogging.

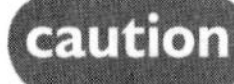
If you want to avoid any technical problems when posting a new entry from the Web browser, you should be connected to the network the whole time you are accessing the blog.

WordPress has introduced small screen–friendly blog pages at m.wordpress.com. Type the address in your phone browser and log on to your account. Click the Post link and write your blog entry. Select the Publish button when you are ready (see Figure 21-11).

Figure 21-11

Blogging in Opera browser.

note *Not all phone browsers come with the advanced features required for accessing a blog. If the built-in browser on your phone doesn't let you post new entries, you might consider Opera browser or Opera Mini. Opera Mini is a free browser that is compatible with a wide variety of phone models. It is available at mini.opera.com.*

Project 22

Create Real Music Ringtones and Video Ringtones

What You'll Need

- **A ringtone editor application or a video ringtone player on your phone**
- **A ringtone and video editor application for your PC**
- **A phone that can play MP3 ringtones or video ringtones**
- **Cost: $10 U.S. for the Vito Ringtone Editor**

  $10 U.S. for the SkyeTones video ringtone player

  $20 U.S. for the Magix Ringtone Maker PC application

Ringtones come in many shapes and sounds. Polyphonic MIDI ringtones are easy to download, and they sound pretty good on the phone. MP3 tunes are as close to real music as a ringtone can get. Surely, there can't be anything better than an MP3 ringtone? Yes, there is. It is called a video ringtone. A video ringtone can perform a song-and-dance act on the screen whenever your phone rings. Read on to discover how you can prepare your phone for the new exciting audio and video ringtones.

In this project, you'll learn to produce ringtones from MP3 songs and from video clips. You may already have plenty of free material available if you have extracted MP3 songs from your CDs. Your DVD library is another source of free ringtone material, especially for creating video ringtones. This project requires an effort from you to install and learn new software, but when you hear and see (see Figure 22-1) the new ringtones on your phone, you'll know it was worth the effort.

This project comprises three sections: creating an MP3 tune on the phone, setting up the phone for video ringtones, and creating audio and video ringtones on a PC. Start from any section you like and continue to other parts when you need the information discussed in them.

Figure 22-1

When your phone rings, the video ringtone springs into action.

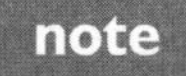

The ringtone editor that we'll be using in the project runs on Windows Mobile phones and the video ringtone player runs on Symbian OS/S60 devices. The Magix Ringtone Maker is a PC application that lets you mix music and video. Its output can be transferred to any phone.

Step 1: Mix MP3 Ringtones on Your Phone

The Vito Ringtone Editor product runs only on Windows Mobile smartphones. However, you can mix your own MP3 ringtones in the Magix Ringtone Maker on a PC and copy the produced ringtones to any phone.

note

Many phone models can play full MP3 tracks as ringtones. When your phone rings, the song is always played from the beginning. If you want a particular segment of a song to play as your ringtone, you have to cut that piece from the song and save it as a ringtone.

1. Install the Ringtone Editor Application

Launch the Web browser on your computer and go to Vito Technology's home page at www.vitotechnology.com. Look for the Ringtone Editor product for Windows Mobile smartphones. You can choose to download a ZIP, EXE, or CAB file.

If you have the phone's data cable at hand and ActiveSync installed on your PC, download the EXE file. When you have saved the file on the hard disk, attach the data cable to the PC and to the phone. Double-click the EXE file to begin installation. You have to confirm the installation on your phone and on your PC.

tip

Downloading the CAB file lets you transfer the application to the phone without having to use a data cable or ActiveSync. Once you have saved the CAB file on the hard disk, you can send it to the phone via Bluetooth. Another option is to copy the CAB file to a memory card using a memory card reader. After inserting the memory card back into the phone, use the File Manager to locate the CAB file and install the application.

2. Open an MP3 Song

After you have installed the application on the phone, you are ready to begin your ringtone production. Go to the Home screen and press START. Find the Ringtone Editor application and launch it. You should see a list of folders (see Figure 22-2). Open the folder where you have saved MP3 tracks.

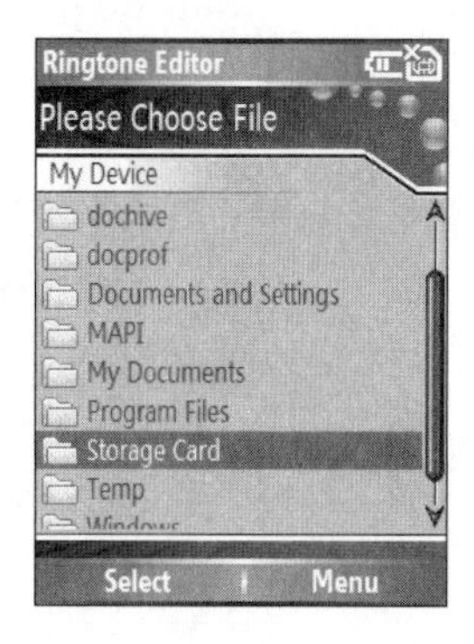

Figure 22-2

Locate your MP3 files for the Vito Ringtone Editor.

In case you don't have any MP3 files saved on your phone, you can download songs from the Internet or from your computer. Follow the instructions in Project 23 for Internet downloads and Project 6 or Project 7 for transferring MP3 tracks from your computer.

If you are using a memory card as your primary storage device, open the Storage Card folder and open the folder where you have saved your MP3 tracks. You may begin your search from the My Documents folder. Inside My Documents, you'll find My Music folder where other applications may have saved songs. When you have discovered the song you want, choose Select.

3. Create a New Ringtone from an MP3 Song

The Ringtone Editor instantly starts to play music when you open a track. Listen to the song and make notes for the start and end points that make a good ringtone. Often, the best candidates are the intro and the chorus parts. Select Cancel if you don't want to listen to the whole song. You can also move the navigation key on the phone key-pad to the left to rewind the track.

Play the song once more. Select Start when the segment for your planned ringtone begins. Select Finish when the ringtone segment ends (see Figure 22-3).

Figure 22-3

Cut the song to create a flashy ringtone.

Don't make your ringtones longer than 30 seconds. Some phones may not play longer tunes at all. Also, MP3 files take quite a lot of memory space.

4. Save the Tune as Your New Ringtone

The Ringtone Editor shows you a number of options (see Figure 22-4) after you have selected Finish. Verify that the piece of music you cut out from the song is what you wanted. Select Play.

Figure 22-4

Assign the tune as your new ringtone.

If the segment isn't exactly what you wanted, give it a new try. Select Re-Edit. You can listen to the song again and cut a new segment with the Start and Finish buttons.

When you are happy with your new tune, select Assign as Ringtone. You can also save the tune for later by choosing Save Ringtone.

It is easy to switch ringtones in the phone Settings as well. Go to the Home screen, select Start | Settings | Sounds. Then press the selection key in the Ring Tone field and pick up a new tune.

Step 2: Set Up Your Phone for Video Ringtones

Video ringtones are an exciting idea. When your phone rings, not only can you hear it, but you can also see it. From now on, phone calls can be all song and dance, or caring words and gestures from loved ones. I'm going to use SkyeTones as the sample product in this project. The software runs on Symbian OS devices.

A number of video ringtone applications are available, but getting some of them to actually play videos can be a challenge. Download the trial version of a video ringtone application before purchasing to make sure it works on your device in the way described in the product description.

1. Capture Video

The easiest way to get video clips into a camera phone is to capture them in the device. Activate the camera, switch on the video mode, and shoot a brief film. Since this video is going to become a ringtone, you need a strong audio signal as well. Phone microphones must be placed close to the audio source to pick up the sound. Avoid making any noise yourself because the microphone will easily pick it up.

tip

When capturing material for your video ringtones, try different camera settings. For example, MP4 video format worked fine for me on one phone model, but 3GP videos didn't play as ringtones at all.

note

Another good source for video clips is movie trailers and music videos you may have saved on your computer. Produced by professionals, they provide a perfect mix of audio and video. You just have to find a segment from a video that makes an impressive ringtone. For this, you need a video editing application. The Magix Ringtone Maker, discussed in this project, will help you with video editing. The products mentioned in Project 24 may be useful as well.

2. Install the Video Ringtone Application

Launch the Web browser on your computer and go to Handango (www.handango.com) Web site. Select your phone model and look for the SkyeTones product. Download a trial version to make sure it works on your device before purchasing. Save the file on the hard disk. Transfer the downloaded SIS file to your phone using a memory card, data cable, or Bluetooth.

In case you transmitted the application to your phone via Bluetooth, open the message in the inbox. The installation process will start. Confirm the requests to install the application.

If you copied the application to the memory card, launch the File Manager (can be found in the Tools folder). Locate the SIS file from the memory card and open it. The installation program will ask for your confirmation before running the installation process. Accept the installation requests.

When the installation is ready, go to the main menu. Open the My Apps (or My Own, Install, or Fun) folder and launch SkyeTones. Select Trial (see Figure 22-5) if you downloaded the trial version. Otherwise, enter the license code you received from the product vendor.

Figure 22-5

Try the product before you buy it to ensure it works on your phone.

SkyeTones

Enter license code.

123

Trial

3. Set Up a Video Ringtone

You can define a video ringtone for a phone profile or for an individual person. If you set the video ringtone for a profile, for example Outdoor, all incoming calls will play the same video when the Outdoor profile is active. You can also assign a specific video ringtone for an individual person, while all other calls play the default ringtone.

For example, you may have a video clip saved on your phone where your partner is in a funny mood or says something nice to you. If you set the clip as your video ringtone, and he or she calls, the video will play on your phone screen. Just remember—if your phone happens to ring in a public place—that the audio plays as well.

Go to the main menu. Open the My Apps (or My Own, Fun, or Install) folder. Launch SkyeTones (see Figure 22-6).

Figure 22-6
SkyeTones brings video ringtones to your phone.

First, let's set a video ringtone for a profile. Open the Options menu in the SkyeTones main screen (see Figure 22-6). Select Profiles and highlight a profile. Open Options and select Personalise. Push the selection key when the Ringing Tone field is highlighted (see Figure 22-7). You will see a list of audio and video clips saved in the phone. Pick up a video from the list. You can identify videos from the names that end with .3gp or .mp4. Select Back and Exit until you are in the SkyeTones main screen again.

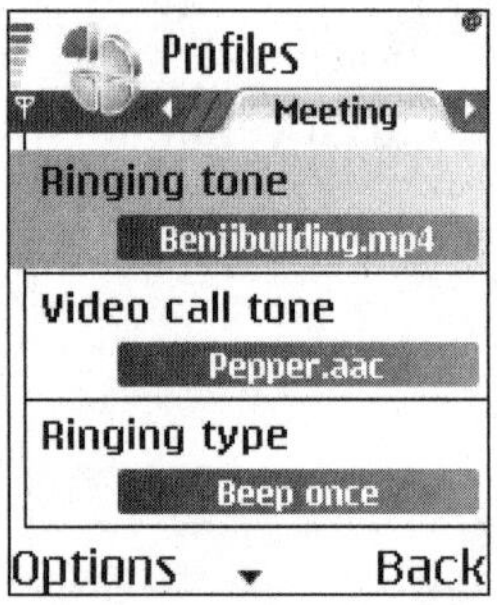

Figure 22-7
Define a video ringtone for a profile.

Second, let's set a video ringtone for that someone special in your life. When you are in the SkyeTones main screen, select Options | Contacts. Find the person you want to assign the dedicated video ringtone to. Open the contact card. Select Options | Ringing Tone (see Figure 22-8). You are viewing a list of audio and video clips saved on the phone. Select a video clip (3GP or MP4). The name of the ringtone you picked up will display in the person's contact card.

tip *You probably don't have a video clip of every person you'd like to. Assuming these individuals have camera phones, request that they capture a few seconds worth of video of themselves. Ask them to send the recording to you via MMS or e-mail. When you receive a video clip in your phone's inbox, save the attachment in the phone memory or on a memory card. Then, follow the instructions in this project for assigning it as a ringtone.*

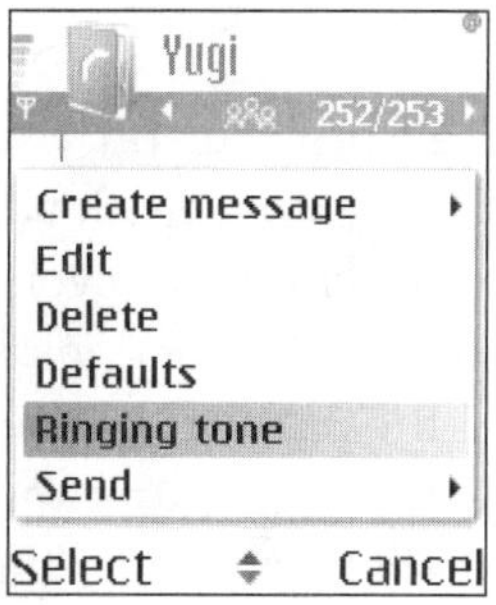

Figure 22-8

Assign a video ringtone for that special person in your life.

Yugi
252/253
Create message
Edit
Delete
Defaults
Ringing tone
Send
Select Cancel

4. Let it Ring

Now, ask someone to call you and watch your new ringtone spring into action (see Figure 22-9). Every time you power on your phone, remember to restart the SkyeTones application. The settings for SkyeTones are retained even when the device is powered off.

Figure 22-9

Is there anything better in modern life than having a video ringtone on the phone?

Step 3: Mix Audio Ringtones and Video Ringtones on Your PC

Not all phones allow you to mix ringtones or video ringtones, but it is possible to create new ringtones on a computer and copy them to your phone. Magix Ringtone Maker allows you to read different types of audio and video files, edit and mix them, and create audio and video ringtones that are compatible with your phone. Other applications that allow you to create ringtones are, for instance, Ringtone Media Studio and Xingtone Ringtone Maker.

1. Install

Launch the Web browser on your PC and point it to Magix (www.magix.net), or Download.com (www.download.com). Look for the Magix Ringtone Maker, but make sure you download the version 2, or newer release. This is because the release 2 includes video editing functions in addition to the audio mixer features.

Download the product on the hard disk. It may take some time to receive the file because of its large size. After the download is finished, double-click the saved EXE file to start installation. Follow the instructions to install.

2. Import a Tune

Launch Ringtone Maker. You should see the main screen which is divided into three panes (see Figure 22-10). Pane 1 lets you search and import tracks, pane 2 is the mixing board, and pane 3 allows you to save your sounds as ringtones.

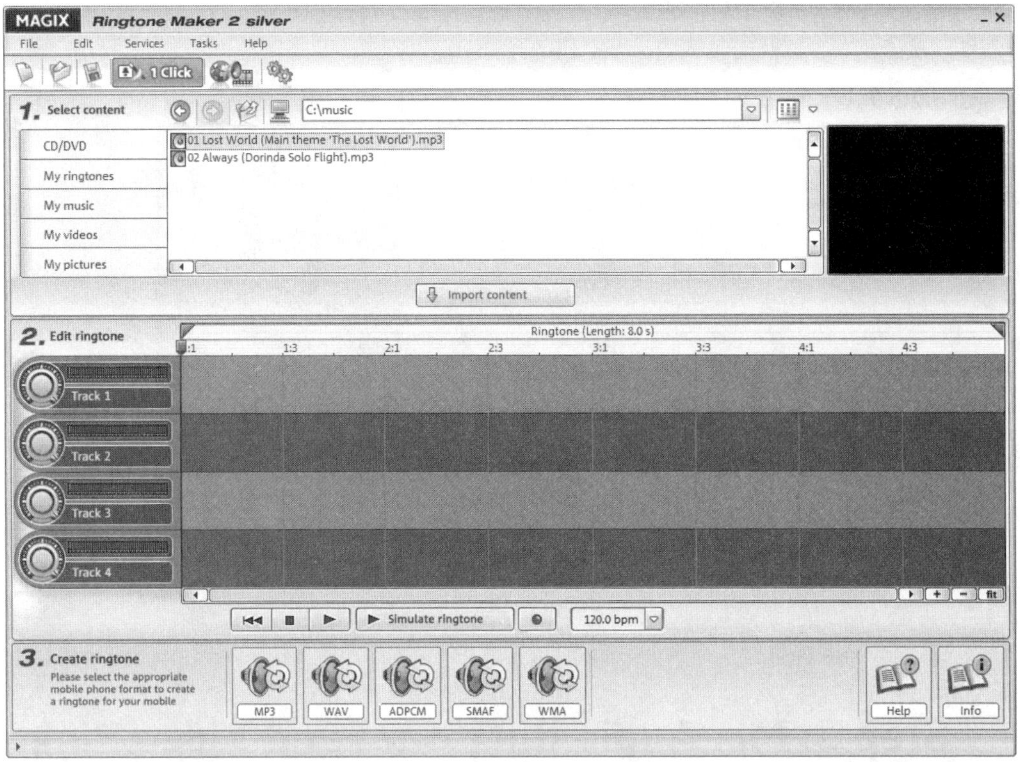

Figure 22-10

The main screen in Magix Ringtone Maker. Pane 1 is for importing items, pane 2 for mixing, and pane 3 for saving ringtones.

Select the button with the computer monitor picture on it. It is located next to the Select Content title at the top of the window. You can import tracks from the hard disk or from CDs and DVDs. Navigate to the folder on the hard disk where you have saved your MP3 or other digital music files. When you find the right folder, Ringtone Maker will list the tracks in the pane at the top.

Highlight a track listed in the folder pane at the top of the program window. Click the Import Content button at the center of the window. The sound wave of the song will show up in the center pane.

3. Create an Audio Ringtone

When you have imported a song into the mixing board, you can edit, merge, and compose a new piece of music. Click the Play button below the sound wave to listen to the track. Follow the time line of the song. Pay attention to the intro and chorus parts that often include catchy segments for tunes. Grab the blue triangle (you can find it above the sound wave) with your mouse and move it to the beginning of your new ringtone. Move the other blue triangle to the end of the segment (see Figure 22-11).

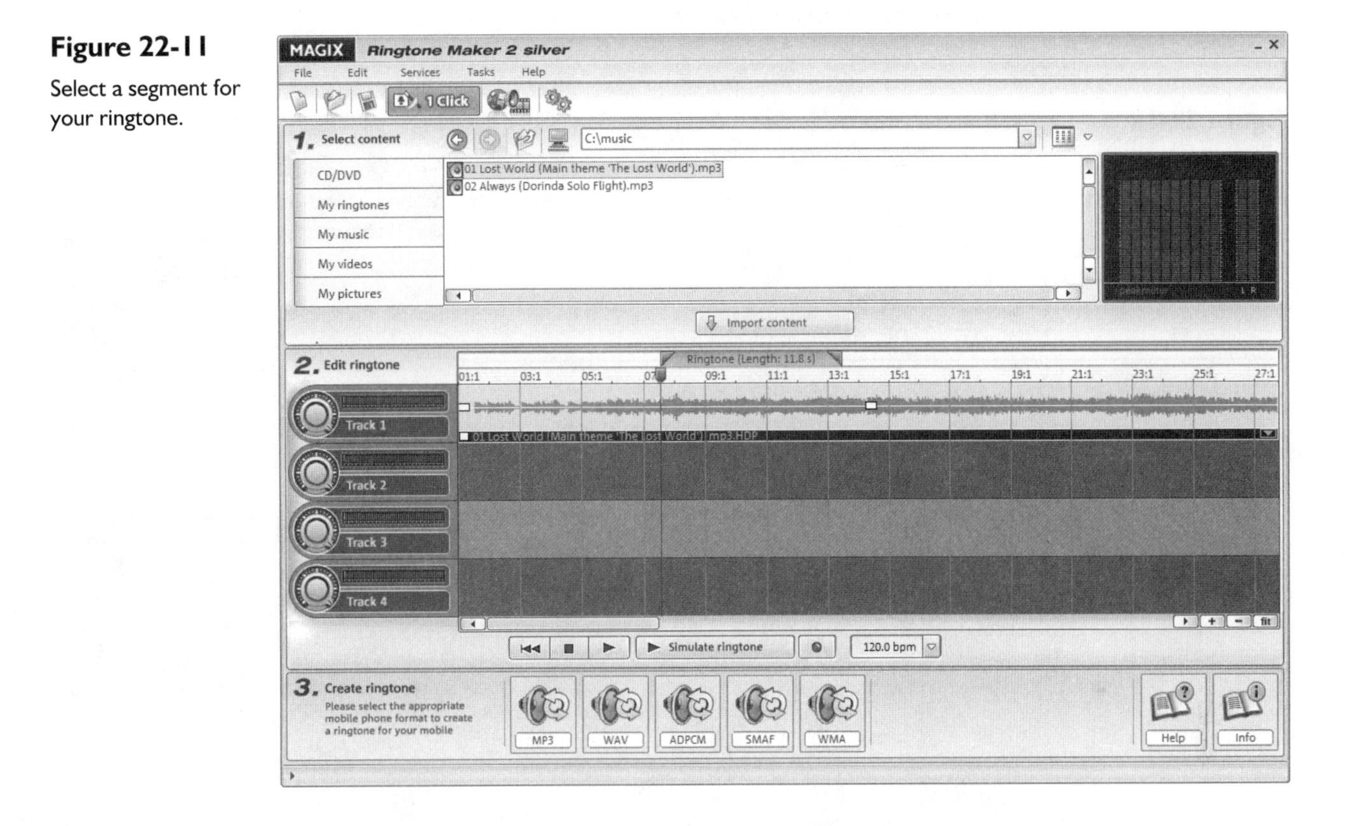

Figure 22-11

Select a segment for your ringtone.

tip *When you begin to work with a new song, it is helpful to be able to view the timing for the whole track. Select the Fit button that's located to the right, just below the sound waves. Once you have identified the whereabouts of the segment you want, you need more detail. Click the button with the plus sign on it to change the time scale.*

Listen to the segment you have selected by clicking the Simulate Ringtone button. If required, make adjustments by moving the triangle signs that mark the beginning and end of the tune.

Don't forget to save your new tune. Later in this project you'll find a section about saving tunes in different file formats.

4. Create a Video Ringtone

You can import videos from your hard disk or movies from CDs and DVDs into Ringtone Maker.

In order to pick up a video file from your computer, click the button with the computer monitor picture on it in Ringtone Maker's main screen. When you have located a video file from your folders or discs, click the Import Content button at the center of the program window. A still image of the video takes one track, and the sound wave reserves another track in the mixing board.

Grab the blue triangle marker with your mouse and move it to the beginning of the segment that will become your new video ringtone. Grab the other blue marker and move it to the end of the new segment (see Figure 22-12). Push the Simulate Ringtone button to watch and listen to the video.

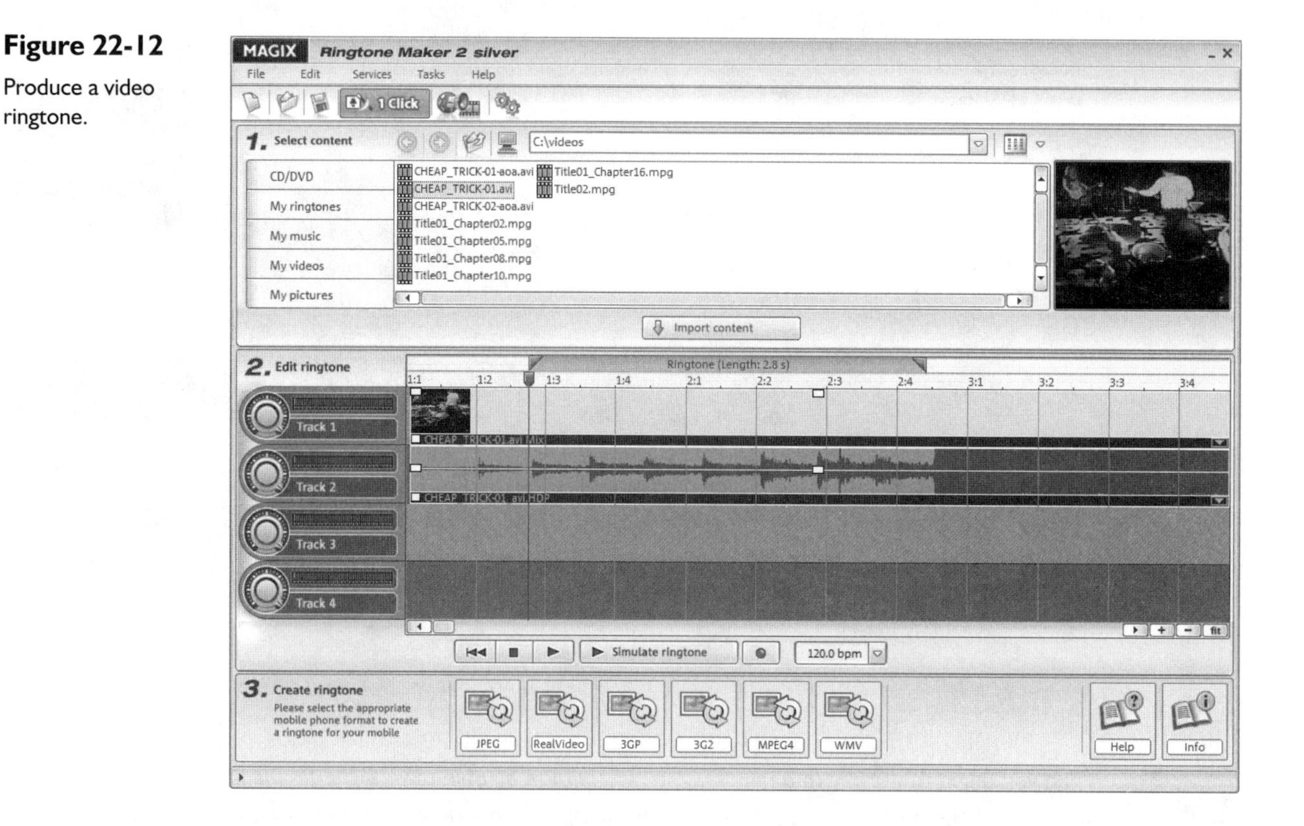

Figure 22-12
Produce a video ringtone.

5. Save the New Ringtone

Ringtone Maker allows you to save your new tune in a variety of formats. If you cut the tune from an MP3 track, choose MP3 as the ringtone format. Click the MP3 button at the bottom of the program window. Select a folder and type a name for the tune in the text box at the top (see Figure 22-13). Click OK to save the file on the hard disk.

tip *If you are unsure which format to choose, click the Info button at the bottom-right corner of the program window. It takes you to a Web page where you can browse a list of phone models and compatible ringtones. You can almost always choose WAV, because it is a universal format recognized by practically all phones, but it takes plenty of storage space.*

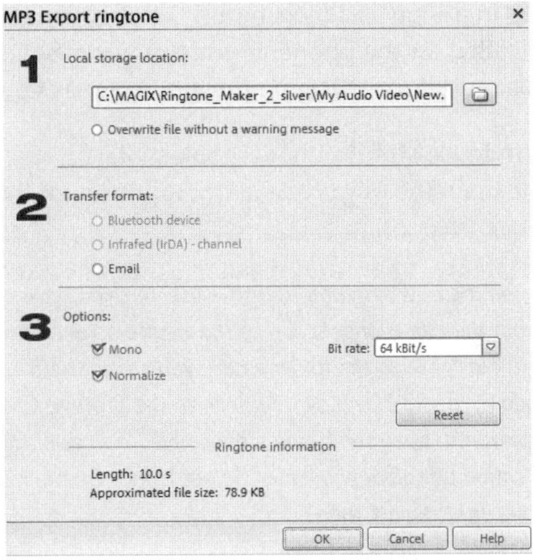

Figure 22-13

Save the new MP3 tune.

Saving a video ringtone is easy, even though there is plenty of choice available in the pane at the bottom of the program window. Click the 3GP button to save the video on the hard disk. Select a folder and enter a name for the clip. Click OK.

Just in case, save the video also in MP4 format by clicking the MPEG4 button. This way, you may try the MP4 file if the 3GP won't play on your phone.

6. Copy the New Ringtone to Your Phone

In case you saved the ringtone file on the hard disk (see Figure 22-13), you can use a memory card or data cable to transfer the tune to your device. Remove the memory card from your phone. Attach a memory card reader to your PC, insert a card into the reader, copy the file to the memory card, and insert it back into the phone.

The Magix Ringtone Maker also lets you transmit the file to your phone via Bluetooth or infrared. Select Bluetooth in the Ringtone Maker's Export window (see Figure 22-13). Pick up your phone from the list of discovered Bluetooth devices. You may have to pair the phone and the computer by typing a code (that you make up) on both devices. When your phone receives the file, open the inbox and save the attachment in the phone or on the memory card.

7. Set the Tune or Video as Your New Ringtone

There's still one more thing to do: set the new tune as your ringtone. You have already saved the file in the phone memory or on the memory card, but you have to tell the device where it can find it.

Symbian OS/S60 Phone To set up an audio ringtone, go to the main menu. Select Tools | Profiles. Choose a profile and select Personalise. Open the Ringing Tone list and pick up your new ringtone.

To assign a video ringtone as your ringer, open the player application you have installed on the phone. If you are using SkyeTones, follow the instructions earlier in this project for changing the video clip as your ringtone.

Windows Mobile Phone Select Start in the Home screen. Select Settings | Sounds and open the Ring Tone list. Scroll through the list for your new ringtone. Select Done when you find it.

tip *If you can't find the new ringtone from the list, but you have saved the tune in your phone, you need help from the File Manager. Go to the Home screen and select Start. Find File Manager and launch it. Navigate to the folder where you saved the file. For example, on my HTC/Qtek device, Windows Media Player syncs music to the Storage Card/Music folder. Highlight the file you have discovered, open the menu, choose File, and select Move To. Navigate to the Smart-phone/Application Data/Sounds folder. Select Done. Go back to the settings, list the ringtones again, and pick up the new tune.*

Project 23

Download Music to Your Phone from the Internet

What You'll Need:

- **An application that lets you download music, or listen to and record from Internet radio stations on the phone**
- **MP3 player software on your phone**
- **A memory card for storing songs**
- **Cost: $20 U.S. for the music recorder application**

Even if you have an extensive selection of your favorite songs stored on your phone, there may be moments when you want to quickly get some new music to go. If you happen to be away from home, you would have to wait until you get back to your computer to download more music. Fortunately, there is a faster way—you can download songs directly from the Internet to your phone.

tip *If you purchase digital music, accept only MP3 tracks. You can play MP3 songs on your phone, computer, iPod, and practically any device that can play digital music. Other music formats are not as widely recognized by digital players, which may lead to compatibility problems. Also, the songs may have been copy-protected. It means that you may not be able to copy and play the purchased music on your other digital devices.*

Two techniques for downloading music from the Internet will be discussed in this project. Choose the one that works the best for you.

- *Accessing an online music store in a dedicated application.* The advantage of this technique is that once the music is downloaded, you don't have to worry about the network connection or the data transfer cost. On the other hand, you must have storage space for music, and your phone must be able to run add-on applications. I will use an application called PeerBox for accessing music stores.

It is also possible to download songs from online music stores using your phone's Internet browser software.

- *Listening to Internet radio.* Accessing Internet radio stations requires an application that can receive shows streamed over the Internet. Some of these applications can record received programs and save them on your phone as MP3 files. It is a handy way of getting new music on your phone, but it requires a continuous connection to the Internet. I will use Resco Pocket Radio and Mobilcast applications for saving music from Internet radio stations on the phone.

note *Downloading music from the Internet to the phone will generate a considerable amount of network traffic. A flat-rate data plan that allows unlimited data transfer, or a data plan with a large number of megabytes per month, is highly recommended. A typical MP3 song is 2–6 MB in size. If you listen to Internet radio, you are constantly streaming data to your phone, and megabytes add up quickly.*

First, let's find out how to download music to the phone from an online store.

Step 1: Download Music from an Online Store

The Beta release of the PeerBox application (which was used in this project) is only available for phones powered by Symbian OS/S60 software. If PeerBox is not yet available for your phone, you can

- Skip to step 2 where you'll find instructions for listening to Internet radio and recording songs from the radio
- Access an online music store using your phone's Web browser (see Figure 23-1)

Figure 23-1
Discovering MP3 songs in a Web browser.

note *Even if your phone is not compatible with PeerBox software, you may download tracks from some online music stores using your phone's Web browser. For example, Peoplesound (www.peoplesound.com) is an online music store that provides promotional MP3 tracks. First, register to the service using the Web browser on your computer. Then, pick up your phone, launch the Web browser, log on to Peoplesound, and find tracks you like. You can download songs from Peoplesound directly to your phone.*

1. Install PeerBox

The easiest way to install PeerBox is to download it directly to the phone from the product's home page. Launch the Web browser on your phone. Type **wap.peerboxmobile .com** in the address bar. When you see the PeerBox home page, confirm the download by clicking the link (see Figure 23-2). The phone's installation routine will start and guide you through the installation process.

Figure 23-2

Download PeerBox to your phone.

http://wap.peerboxmobile.com/
Use your phone to access
the richest catalog on earth
Download PeerBox
Options Close

tip *If you don't manage to download the application to your phone from the Web page, you can download it to your computer. Launch the Web browser on your computer and go to the Web page www.peerboxmobile.com. Click the Download link. Download the product version that is compatible with your phone. Save the file on the hard disk. Use a memory card, Bluetooth, or data cable to transfer the application to your phone.*

2. Find a Song

Go to the main menu and open the My Apps (or Install, My Own, or Fun) folder. Launch PeerBox. The product provides three ways for discovering songs: search, browse an online catalog, and song identification.

Select the Catalog icon (see Figure 23-3). You can browse the music catalog by artist, genre, or other criteria.

Figure 23-3

Browse the music catalog in the PeerBox.

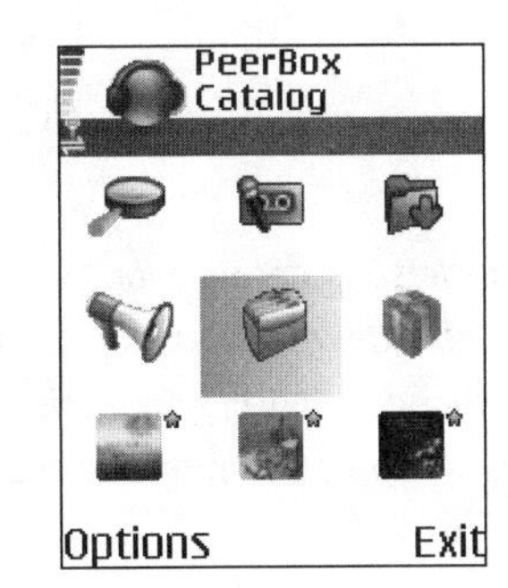

tip *PeerBox can tell you the name of a song after it has listened to a short audio sample. When you hear a song, for example, on radio or on TV, let PeerBox record a short segment of the music. Do as follows. Select Identify in the PeerBox main menu. Keep the phone microphone close to the audio source while PeerBox is recording the sample. In a moment, it should tell you the name of the song.*

3. Download a Song

When you have discovered a song you like from the PeerBox catalog, you can view its details. Download the song by clicking on the Download or Buy button (see Figure 23-4). Some songs are available for free (you'll see Download button at the bottom of the screen) and some have a price tag (Buy button is displayed). PeerBox will ask for your approval before it initializes a download for a song that you have to pay for. The song is charged on your phone bill.

Figure 23-4

Download MP3 tracks to your phone from the PeerBox store.

The download will begin in the background. If you want to follow its progress, click the Back button until you are back in the main menu. Select Downloads for viewing ongoing and past downloads (see Figure 23-5).

Figure 23-5

View the progress of your download.

Earlier, when you viewed the song details, you may have spotted an indication for the download speed of the track. If the speed was listed as fast, it depends on your network connection how quickly the song will download. A Wi-Fi connection can transfer a full MP3 song in a minute, whereas it may take longer than 10 minutes to receive the same track over a GPRS connection.

4. Listen to Downloaded Songs

When the download is completed, you can listen to the song on your phone. Go to the PeerBox main menu and select Downloads. Highlight the song you want to play and press the selection key.

Step 2: Listen to Internet Radio and Save Songs on Your Phone

There are two types of radio receivers available for phones:

- FM tuners that are built into the device hardware. The headset cable doubles as an antenna that picks up FM radio waves. A piece of software that runs on the phone lets you tune into radio stations.
- Internet radio applications that make it possible to listen to radio stations that broadcast over the Internet. As long as your phone is connected to a mobile network (preferably EDGE, EV-DO, UMTS, or HSDPA) or to a Wi-Fi access point, you can listen to Internet radio stations.

We are going to focus on accessing Internet radio stations in this project.

1. Install the Internet Radio Application on Your Phone

You can't add a hardware radio to your phone as an add-on component, but you can install an Internet radio application. We are going to tune to Internet radio stations using S60 Internet Radio (free application) on a Symbian OS/S60 phone, Resco Pocket Radio ($19.95 US) on a Windows Mobile phone, and Mobilcast (free application) on other phones.

Symbian OS/S60 Phone Start by launching the Web browser on your computer. Go to the Web page opensource.nokia.com/projects/s60internetradio. Download the version of the S60 Internet Radio application that is compatible with your phone. Save the file on the hard disk.

Transfer the application to your device using a memory card, Bluetooth, or data cable. If you have a data cable and PC Suite software, PC Suite can install the application for you. You can find detailed instructions for copying information to your phone via a memory card or via Bluetooth in Projects 6 and 7.

Once the application is saved on your phone, you have to install it. If you sent it via Bluetooth, open the message from the inbox, and the installation routine will take care of the rest. In case you copied the new piece of software to a memory card, launch File Manager (File Mgr). Find the application from the memory card and open it. The installation process will start. Follow the instructions until the installation is completed.

Windows Mobile Phone Launch the Web browser on your computer and go to an online software store, such as Smartphone.net, www.smartphone.net. Search for the Resco Pocket Radio for Smartphone application. When you find it, download it and save it on the hard disk.

Hook up the data cable from your phone to your PC. Double-click the file you downloaded to initiate installation. Follow the instructions on your PC and on the phone until the installation is finished.

Other Phones Launch the Web browser on your computer. Go to the Mobilcast registration page at www.melodeo.com. Check your phone's compatibility with Mobilcast. If it is compatible, register with Mobilcast. During the registration process you will receive a text message with a link to the downloadable file, or you will see the download link that you can enter into your phone's Internet browser. I got the application from zeus.melodeo.net/mc, but you should verify the address during your registration. After the download, installation should begin automatically and all you have to do is to confirm that you really want to install it.

2. Tune into an Internet Radio Station

When you are listening to AM, FM, or satellite radio, you have to be able to pick up a station's radio signal. But when you are listening to the Internet radio, you have to be connected to a mobile network, because the shows are streamed to the data communication channel on your phone.

Symbian OS/S60 Phone You can find the S60 Internet Radio in the My Apps (or Fun, My Own, or Install) folder. Launch the application.

First, open the Options menu and make sure you are listening to radio and not to MP3 songs saved on your phone. Select Change Mode and choose Internet Radio. Reopen Options and select Stations. A few radio stations have been preset in the application. Pick up one (see Figure 23-6). The application will connect to the Internet for receiving a music stream from a radio station. Hook up your headset and enjoy.

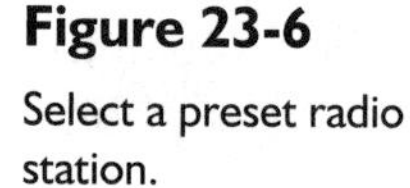

Figure 23-6

Select a preset radio station.

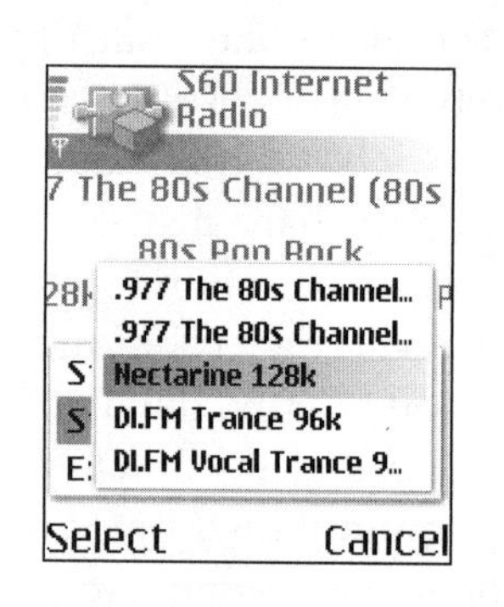

tip *You can add more radio stations to the S60 Internet Radio by creating a file that specifies the details of a broadcast. Visit the Web page of your favorite radio station and look for the Shoutcast (.pls) file. An extensive list of radio stations with PLS files is available at www.shoutcast.com. When you find a station you like, save the PLS information as a text file. Copy the file to your phone and place it in the Shoutcast folder on the drive where you installed the S60 Internet Radio application.*

Windows Mobile Phone Go to the Home screen and select Start. Find Resco Radio from the menu and launch it. If this is the first time you are using Resco Radio, you have to specify the speed of your network connection (see Figure 23-7). Check the Slow box if you have GPRS connectivity on your phone. Check the Fast box if your mobile device features EDGE, EV-DO, HSDPA, UMTS, or Wi-Fi.

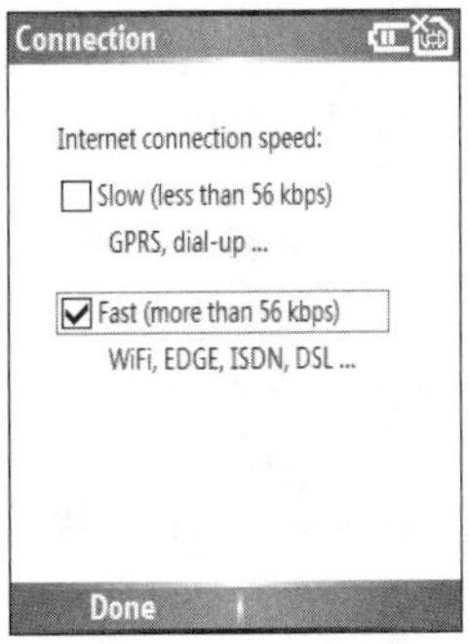

Figure 23-7

Specify your phone's communication capability.

Next, you'll see the Home screen of the radio application. Ten preset radio stations are available on the main screen. Move the selection key left or right to pick up a station. You can also choose a station by selecting Action, choosing Preset, and selecting a station (see Figure 23-8).

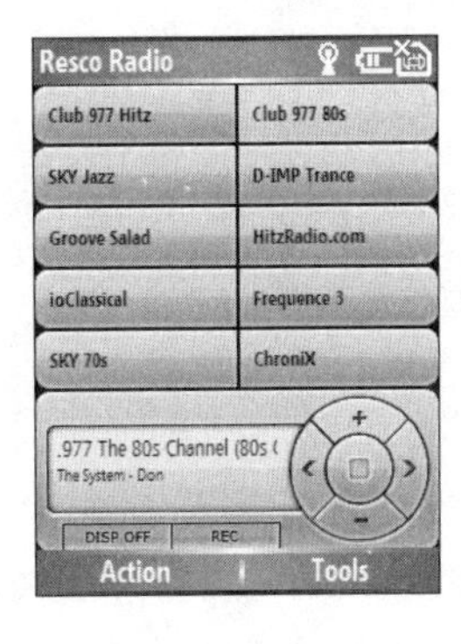

Figure 23-8

Tune into an Internet radio station.

Other Phones Open your phone menu, find Mobilcast, and launch it.

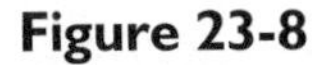

Mobilcast is a Java application. If you have problems finding Mobilcast from your phone's menu system, the phone may have saved the application in a special place for Java programs. Look for Java, Midlet Manager, or the new applications section on your phone for Mobilcast.

When Mobilcast is starting, it refreshes its program directory. It takes a while before you can see its main program categories. Navigate the online catalog of podcasts and music until you find the piece of music or podcast you like (see Figure 23-9). Select Listen Now.

Figure 23-9

Listen to music in Mobilcast.

note *If your phone can't play music without interruptions, but the audio stream frequently stops and restarts, your network connection may be too slow for listening to radio. Sometimes phones even with fast network connectivity, like EDGE, EV-DO, HSDPA, or UMTS, drop into a slow network mode, such as GPRS, if they are out of the fast network coverage. You should stay put in the fast network coverage area if you want uninterrupted transmission. Another solution is to find radio stations that broadcast music in lower quality. Lower audio quality generates less traffic into the network, and it streams through slower connections uninterrupted. Look for Internet radio stations that broadcast in 64 Kbps or lower audio quality for listening over a GPRS connection.*

3. Record a Song

Some applications let you record tracks you hear from the Internet radio stations. You may record as many songs as the storage space on your phone allows.

Symbian OS/S60 Phone The S60 Internet Radio application can't record music, but you can use, for example, PeerBox to save songs from the Internet on your phone.

Windows Mobile Phone Before you record anything, make sure that the music will be saved on a memory card. When you are in the Resco Radio main screen, select Tools | Settings | Buffering & Recording. Go to "Location for Recordings" and select Storage Card (Figure 23-10). Also check the box for "Ask for File Name after Recording."

Figure 23-10

Save recordings on a memory card.

General
Prebuffering with length:
5 seconds
Location for recordings:
Storage Card
Ask for file name after recording
Done | Menu

When you are listening to music in Resco Radio and you hear a song you want to save, select Action | Record/Stop. The recording will start and go on until you stop it by selecting Record/Stop again (see Figure 23-11). Once you end the recording, you can give a name to the tune. Resco Radio automatically records music in MP3 format.

Figure 23-11

Record songs from the radio.

Resco Radio
Club 977 Hitz
Club 977 80s
SKY Jazz
D-IMP Trance
Groove Salad
HitzRadio.com
1 Play/Stop
Frequence 3
2 Preset
ChroniX
3 Prev preset
4 Next preset
All
5 Record/Stop
Action | Tools

Other Phones Mobilcast lets you save some programs. If you can spot the title "Save For Later" in the program screen (see Figure 23-12), you can save the show or piece of music on your phone.

Figure 23-12

Save the show on your phone.

note *Check out the music download services provided by your network operator as well. Many mobile network operators offer music downloads for their subscribers who have a compatible phone with the service. The advantage of downloading music from your service provider's online store is that everything has been made as easy as possible. You don't have to worry about installing applications or finding songs from a variety of sources. The downside is that songs purchased from these stores can be pricey and they typically come with copy-protection restrictions. For example, you may not be able to play the downloaded songs on any other device than on the phone where you downloaded them.*

Project 24

Watch Movies and Videos on Your Phone

What You'll Need

- **A piece of software that can turn movies into 3GP videos on your computer**
- **A DVD copy application (if you want to watch your DVDs on your phone)**
- **A memory card or hard disk for storing movies on your phone**
- **A media player software that can play videos on your phone. Most smartphones that are powered by Windows Mobile, Symbian OS, or Palm OS operating system software and most camera phones come with an application that can play videos**
- **Cost: $20–$40 U.S. for the DVD copy application and $20–$35 U.S. for a 1 GB memory card**

It's amazing what you can do with a phone: watch live TV, listen to music, take photographs, create new tunes, capture home videos, not to mention accessing e-mail and the Internet. Surely no one would want to watch movies on a phone? After all, the screen is small and the battery may quickly dry up. But on second thought, why not? You never know when you have an hour and a half of time to kill and you are still three episodes behind on *Oprah* or the *Simpsons*.

The whole process of transforming a movie into a video that you can enjoy on your phone consists of five steps:

1. Discover a movie.
2. Save the video on your computer's hard disk.
3. Convert the video into a digital format that's compatible with your phone.
4. Copy the material to your phone.
5. Watch it (see Figure 24-1).

Figure 24-1
Watch movie trailers, music videos, and other fun stuff on your phone.

This project requires some experience with computers. A basic understanding of video formats and resolutions will help, too. This is because you may need to try out different settings in video conversion programs. Once you have successfully completed the project and discovered the video transformation process that works the best for your phone, the next video transfer will be easy.

The reward for your effort is that you can watch movies or TV shows on your phone anywhere you want, even without a network connection.

Step 1: Discover Movies

Three easily accessible sources for finding movies are your own DVD collection, your digital video recorder (if you have, for example, a TV tuner hooked up to your computer that can record TV shows on the hard disk), and the Internet.

If you want to copy a movie from a DVD, go to step 2 which explains how to extract movies from DVDs. If you have recorded TV shows on a hard disk, you can skip to step 4 which explains how to make phone-compatible videos.

In this project, we are going to download movies, TV shows, and videos from the Open Media Network (www.omn.org). Other Web sites that let you download free movie trailers and music videos are, for instance, Divx (community.divx.com/movies) and Allocine.com (www.allocine.fr/video/podcast.html).

note *Purchasing or renting movies online in order to download them to your home computer is becoming popular as well. Check out, for example, Cinemanow (www.cinemanow.com) or Guba (www.guba.com) for new movies. Downloading a movie from an online store may require that you are connected to the Internet in a country where the store is allowed to distribute movies. In addition, digital copy-protection may complicate things. Amazon Unbox Video Downloads provides you with a compact copy of the video in copy-protected WMV format (Windows Media Video which is labeled as PlaysForSure when copy-protected) that you can transfer to your mobile device. Some new high-end phone models can play these videos. If the online store doesn't provide the movie in a format that is already compatible with your phone, you can't convert the movie because of the copy-protection.*

You may also find an easy access to video downloads from your network operator's Web pages. Some network operators provide music downloads and video downloads for their subscribers. If your service provider has an online store and it is possible to access the store on the phone, check if you can download or view videos without having to go through the video conversion process.

Step 2: Extract Movies from Your DVDs

If you have discovered a DVD that you'd like to watch on your phone, you have to copy it to your computer. Extract selected chapters or the whole movie from the DVD and save the material on the hard disk. To do this, you need a DVD copy application. A wide selection of software tools are available; we are going to use a product called Easy DVD Rip.

In some countries, copying or extracting material from copy-protected DVDs is not allowed. Check the DVD for potential restrictions and the local regulations before you attempt to copy your DVDs.

1. Install a DVD Copy Application on Your Computer

Go to, for example, to Download.com (www.download.com) and search for the Easy DVD Rip product. Download and save the software on your PC's hard disk. Follow the instructions for installing the application.

2. Copy a DVD

Insert a DVD disc into the disc drive. Launch Easy DVD Rip. Choose Convert Chapter in the program's main menu. Select a chapter that you want to copy by checking the box in front of the title. Highlight the title, and you'll see the options (see Figure 24-2).

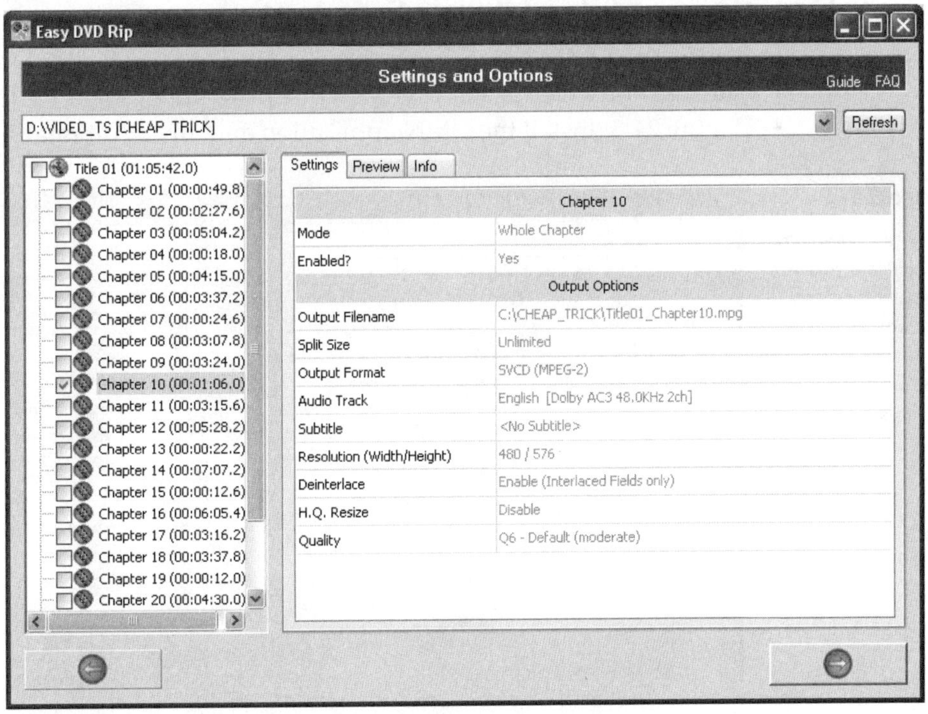

Figure 24-2

Choose the format for the video you are going to extract from a DVD.

Click Output Filename to change the folder where the video will be saved. It doesn't matter where the folder is, as long as you remember its name when you convert the video for your phone.

Make sure the output format is VCD MPEG1. Click on the right arrow at the bottom of the application window. View the top-left corner to ensure the conversion information is correct. Click the GO button at the bottom of the window.

tip *Easy DVD Rip can create MPEG1, MPEG2, and AVI (Divx, Xvid) files from a DVD. If you intend to watch the movie on your phone only, choose a video format that takes the minimum amount of storage space. Select MPEG1 format for videos that you are going watch on your phone. To watch videos on your computer, select MPEG2 or AVI. Later, if you want to watch these movies on your phone, you can change their format to 3GP or MP4.*

Once you have started the video conversion process, it's not worth waiting by the PC. The process is so slow that you can leave your PC running while you go shopping (or even while you sleep if you have a slow PC and a long video to rip). You may continue to step 4 if you want to copy this video to your phone.

Step 3: Download Movies, TV Shows, and Video Clips from the Internet

There are many Web pages on the Internet where you can watch movies, TV shows, and video clips, but not that many that let you download videos. A good place to get started with video downloads is the Open Media Network (OMN) Web site.

1. Download a Video Library Software

Go to the OMN home page at www.omn.org. Click the Download button to get the free software. It will become your program guide and download manager (see Figure 24-3. Install the OMN application and any additional software if requested.

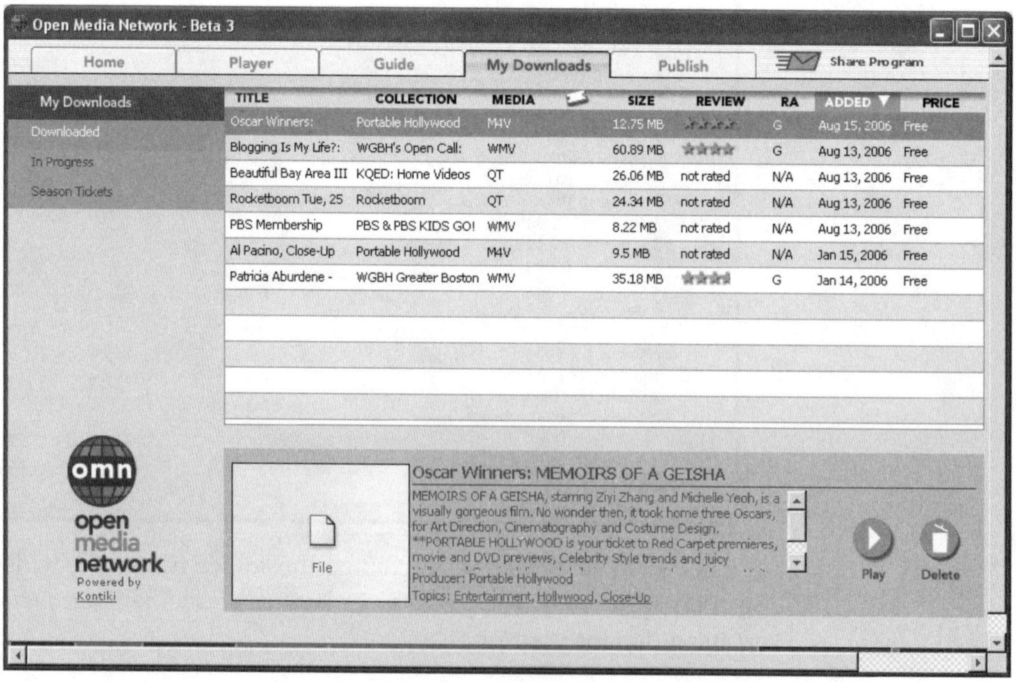

Figure 24-3
The OMN download manager lets you browse and search for videos, podcasts, MP3 music, video blogs, and TV programs.

Before downloading anything, check the folder where you are going to save the material. At the bottom of the OMN window, click Settings. Select Delivery Settings in the left frame. You can see the path to Deliveries Folder. Change this, for example, to a folder that Windows Media Player monitors for new items so that they are automatically added to the media library. In any case, remember this folder because you will need it when converting videos for the phone.

2. Download a Video

Once you have installed the OMN application, you can activate it by clicking the small OMN icon in the Windows Start toolbar. When the program window has opened up, click the Guide tab. You can browse different types of video material. Some items are free, and some items have a price tag.

Click the Get Item button (see Figure 24-4) when you have found something worth watching. The download will start in the background while you can do something else.

Figure 24-4

Start the download in OMN.

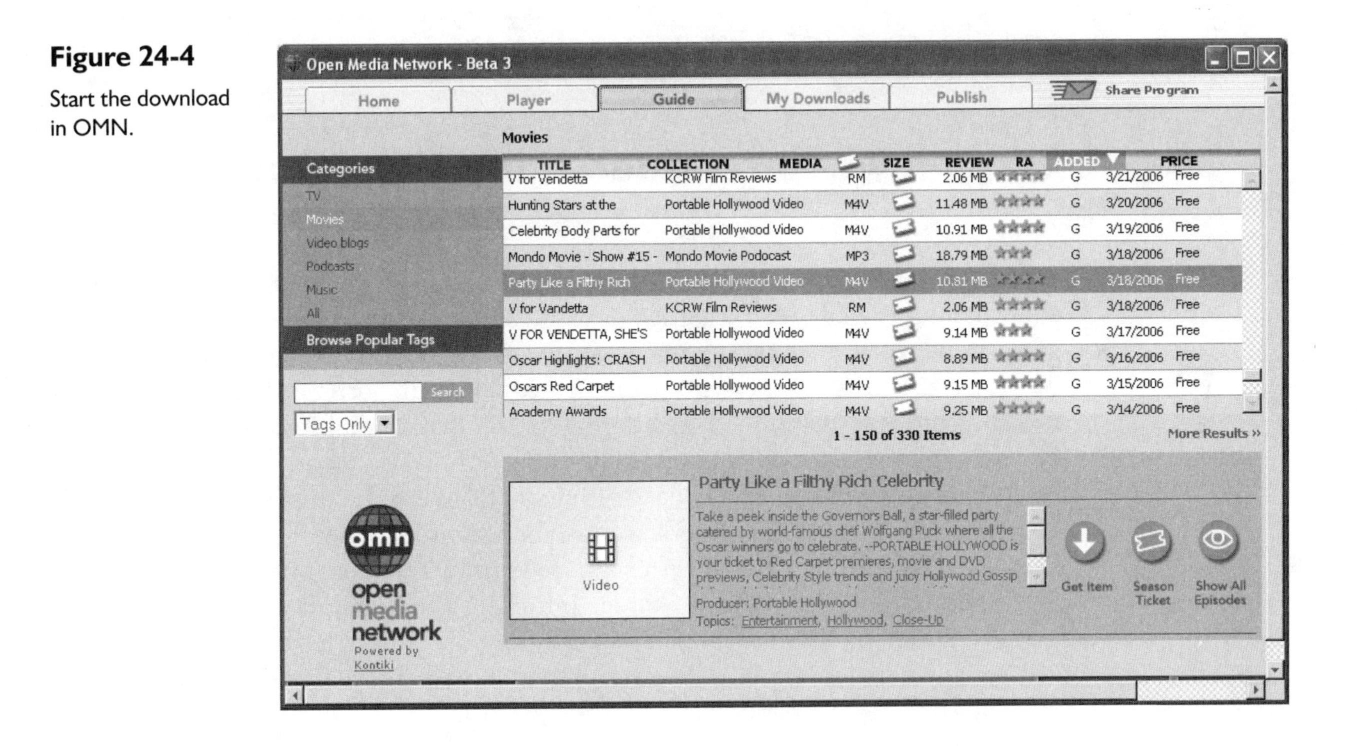

You can follow the progress of your downloads when you click My Downloads at the top of the window. List videos you already have retrieved from OMN by clicking Downloaded on the left. The OMN comes with a video player as well (see Figure 24-5). When the download is finished, highlight the video you want to watch and click Play at the bottom-right corner.

Figure 24-5

If you can't wait until you have converted the movie and copied it to phone, watch it on the PC.

Step 4: Turn Movies into Video Clips That Your Phone Can Play

It doesn't matter whether you downloaded a movie from the Internet, recorded a TV show in your digital video recorder, or extracted a clip from your own DVD, it is not likely to be compatible with your phone. That's why you need a video conversion program that turns videos into a phone-friendly format. Fortunately, the telecommunications industry has agreed on a digital format that all multimedia phones should be able to play. The format is called 3GP.

We are going to use the Nokia Multimedia Converter in this project. Other tools available for 3GP video conversion are, for example, Video Converter 2005 and WinAVI. The Multimedia Converter is a free product that only does one thing: converts a myriad of digital video formats into 3GP. The brand of your phone doesn't matter, as long as it can play 3GP. The Multimedia Converter can turn AVI, Divx, MOV (Quicktime), M4V, MP4, MPG, and Xvid movies into 3GP videos, but it doesn't recognize the WMV (Windows Media Video) format.

1. Install a Video Conversion Application

Visit the Forum Nokia Web site at www.forum.nokia.com and search for the Nokia Multimedia Converter. Download and save the application on your computer's hard disk. Follow the instructions for installing the product on your PC.

2. Turn a Movie into a 3GP Video

Launch the Multimedia Converter. When you see the product's main screen, open the File menu and select Open. Find the video you had saved earlier on the hard disk.

tip *To convert a Divx movie into 3GP, you have to rename the original Divx video file from .divx to .avi for the Nokia Multimedia Converter to recognize it.*

To make sure that you are going to convert the right video, click the Play button at the lower half of the window (see Figure 24-6). Click the Convert button when you are ready to turn the movie into 3GP.

Figure 24-6
Convert a movie into 3GP.

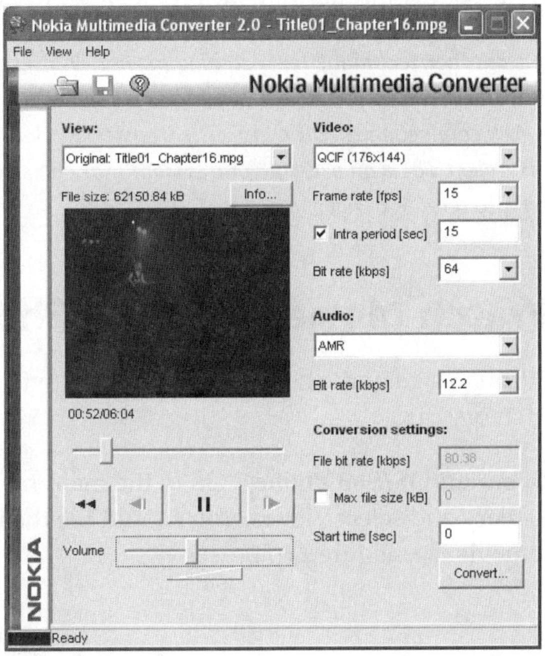

note *The video and audio quality will degrade in the conversion process. Don't expect to see flawless images and to hear hi-fi sound in 3GP. If you are familiar with your phone's capabilities, you can use video formats that allow better video and audio quality than 3GP. For example, if your phone is powered by the Windows Mobile operating system, you can play WMV videos on the device as well.*

3. Save the Video

When the conversion is over, you'll see the program's main screen again. Under the View label, you should see the text "Converted." You can also see the new file size in kilobytes. It helps you estimate if you can fit the item into the phone memory or memory card.

The new video hasn't been saved yet. Open the File menu and select Save. Save the video in a folder where you can find it later in this project.

tip

For minimizing the required storage space, it is possible to further compress the size of the 3GP file by reducing the frame rate and resolution of the video. Simply change a lower frame rate and bit rate for the video and audio in the Multimedia Converter.

Step 5: Copy Movies to Your Phone

Now, you have a 3GP video stored on your computer, but you want to watch it on your phone. Let's copy the item to your phone.

A memory card and data cable are the fastest and most reliable methods for transferring large video files to your phone. For more details about copying information from your computer to your phone, refer to Project 6.

note

Streaming is another technique for watching movies. With streaming, you don't have to download the movie to the phone at all. It requires a fast (in practice, UMTS, EV-DO, HSDPA, or Wi-Fi) network connection that is constantly transferring data during the movie. The Orb server, discussed in Project 18, is an example of a system that can stream TV, videos, and music from your home PC to remote devices.

Step 6: Watch Movies on Your Phone

At this stage, you should have a 3GP movie saved in your phone memory or on a memory card.

Symbian OS/S60 Phone Go to the main menu and open the Media folder. Launch Gallery and select Videos. You should see the latest videos at the top of the list. Select a movie (see Figure 24-7).

Figure 24-7

Watch the downloaded movie.

Windows Mobile Phone In the Home screen, select Start. Find Windows Media Player (or Video Player) and launch it. Open the menu and choose Update Library. If you can't see the Update Library entry at all, select Library | Update Library. Also, make sure that you are listing items stored on a memory card by opening the menu and selecting Library | Storage Card.

Select My Videos | All Videos. Open your new video and the Media Player will play it (see Figure 24-8).

Figure 24-8

Watch the downloaded music video.

tip *If your phone is running Windows Mobile 2003 or older software, the easiest way to find the video from the memory card is to use File Manager. Launch File Manager and navigate to the Storage Card folder where you copied the video. Open the item, and it will play in the Video Player.*

Other Phones Find the video player application from the phone menu. Launch the player and locate the downloaded video from the memory card. Open the video to watch it.

If you thought that the many formats of digital music were unnecessary, digital video has even more variety. Abbreviations like AVI, Divx, M4V, MOV, MPEG, MPEG2, MP4, WMV, Xvid, and many others become familiar for anyone who is downloading videos from the Internet. Each different movie format utilizes a different image compression and encoding technology. Some technologies produce large files and high image quality. Others, like the 3GP, maximally compress the video and audio to produce a small file, but of low quality.

3GP is almost an exception in the video format jungle. The telecommunications industry has managed to agree to a standard that works across devices and networks. MMS (Multimedia messaging), which makes it possible to send media from one phone to another, specifies 3GP as its video format. The 3GP has been defined by the Third Generation Partnership Project 2 (3GPP), which is one of the organizations developing the 3G network standards.

Index

D

E

F

G

H

I

T

U

V

W

X

Y

Z